Materials Horizons: From Nature to Nanomaterials

Series Editor

Vijay Kumar Thakur, School of Aerospace, Transport and Manufacturing, Cranfield University, Cranfield, UK

Materials are an indispensable part of human civilization since the inception of life on earth. With the passage of time, innumerable new materials have been explored as well as developed and the search for new innovative materials continues briskly. Keeping in mind the immense perspectives of various classes of materials, this series aims at providing a comprehensive collection of works across the breadth of materials research at cutting-edge interface of materials science with physics, chemistry, biology and engineering.

This series covers a galaxy of materials ranging from natural materials to nanomaterials. Some of the topics include but not limited to: biological materials, biomimetic materials, ceramics, composites, coatings, functional materials, glasses, inorganic materials, inorganic-organic hybrids, metals, membranes, magnetic materials, manufacturing of materials, nanomaterials, organic materials and pigments to name a few. The series provides most timely and comprehensive information on advanced synthesis, processing, characterization, manufacturing and applications in a broad range of interdisciplinary fields in science, engineering and technology.

This series accepts both authored and edited works, including textbooks, monographs, reference works, and professional books. The books in this series will provide a deep insight into the state-of-art of Materials Horizons and serve students, academic, government and industrial scientists involved in all aspects of materials research.

More information about this series at http://www.springer.com/series/16122

Bangalore Sridharan Sugun
Editor

Practical Approach to 3D Weaving

Springer

Editor
Bangalore Sridharan Sugun
CSIR—National Aerospace Laboratories
Centre for Societal Missions and Special Technologies
Bengaluru, Karnataka, India

ISSN 2524-5384 ISSN 2524-5392 (electronic)
Materials Horizons: From Nature to Nanomaterials
ISBN 978-3-030-80423-7 ISBN 978-3-030-80424-4 (eBook)
https://doi.org/10.1007/978-3-030-80424-4

This Springer imprint is published by the registered company Springer Nature Switzerland AG
The registered company address is: Gewerbestrasse 11, 6330 Cham, Switzerland

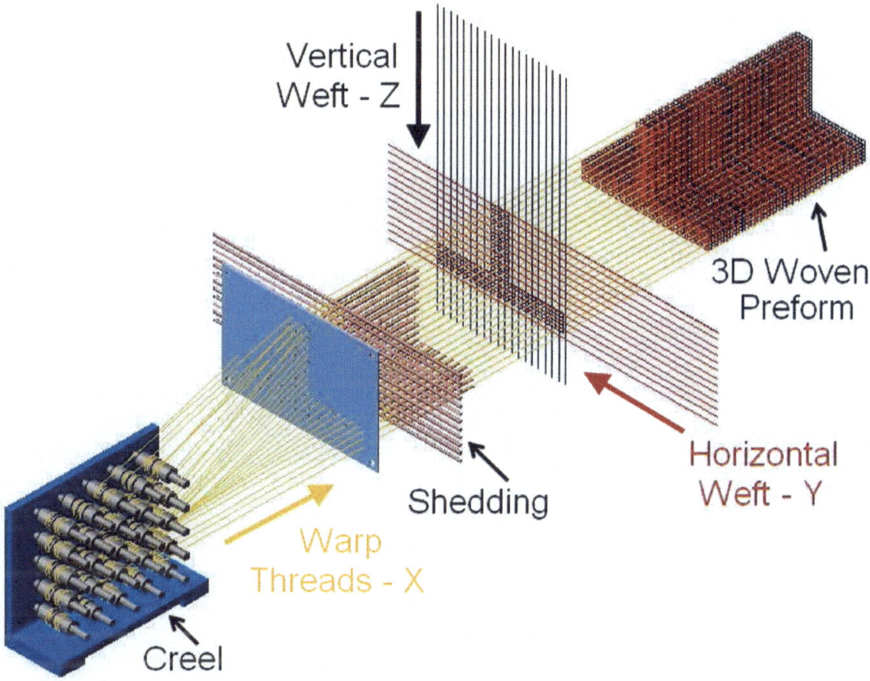

Vertical
Weft - Z
3D Woven
Preform
Horizontal
Weft - Y
Shedding
Warp
Threads - X
Creel

Dedicated to

Jagadaachaarya Shree Bhagavadh-Ramanujaacharya

With blessings from

"Srimathe Sri Lakshmi Hayagreeva Divya Paduka Sevaka Srimadh Abhinava Vaagheesha Brahmathanthra Swatantra Parakala Maha Deshikan"

Acknowledgements

Dr. Raja Manuri Venkata Gopala Krishna Rao, Former Head, Fibre Reinforced Plastics—Pilot plant, Council of Scientific & Industrial Research—National Aerospace Laboratories (CSIR-NAL), for imbibing the keenness to pursue Applied Research against all odds.

Dr. Ramesh Sundaram, Head, Centre for Societal Missions & Special Technologies, CSIR-NAL, for his practical approach and enabling the work on technology development.

Professor Prasad Potluri, Professor of Robotics and Textile Composites, Director of Research, Northwest Composites Centre/NCCEF, University of Manchester, United Kingdom, for his guidance to the editor during Raman Research Fellowship in 2015.

Former Directors of CSIR-NAL Dr. T S Prahalad, DR KN Raju, Dr, B R Pai, Dr A R Upadhya, Mr. Shyam Chetty for their support in fostering research during their respective tenures.

Current Director of CSIR-NAL Shri Jitendra J Jadhav for his support towards technology development.

Several Project students who worked in various phases of 3D weaving Technology development at CSIR-NAL, whose reports have been acknowledged at the end of each chapter.

Special thanks to D N Sandeep, (formerly Project assistant) for his contributions on the shop floor.

Aeronautical Research & Developmental Board, India, for their support in fostering applied research within the country.

Advanced Composites Division and Aircraft Prototype Manufacturing Facility at NAL for their support in machine development.

Colleagues of Centre of Societal Missions & Special Technologies for their direct or indirect contribution.

Special thanks to G. M. Kamalakannan, Senior Principal Scientist, CSIR-NAL and his team for the timely help in reviving and repairing the Electronics part of the dobby loom.

My Guruji Achaarya L S Bharatha Narayana, Wife- Lalitha N S and Son- Rakshan S, other family members and friends for their support and motivation in my research career.

Indian Textile industries viz., M/S Kale Texnique, M/S Arun fabrics, M/s Intertek India Pvt. Ltd., for their support in Technology Development.

Principal and staff of Govt. Sri Krishna Rajendra Silver Jubilee Technological Institute, K R Circle , Bangalore.

Executive Summary

Two-dimensional Weaving is a well evolved, established and versatile technology with primary use in apparel, but encompassing a whole spectrum of applications such as technical fabrics for aerospace and defence, bandage materials and gauges for medical textiles, reinforcement fabrics for the construction industry and reinforcement members in automobile parts. With developments in the strategic sector, transiting to the third dimension became a pre-requisite for some of the applications, as the parts are required to accommodate the multi-dimensional stresses during their usage. An increasing number of researchers, have referred to the third dimension in reinforcements as a "simple integrated buildup of thickness by any means- However, the crux has been aimed at,interlinking the layers together'. Some of the simplest approaches revolved around utilizing stitching, tufting, Z-pinning and related methods. The advantage of these approaches were that, they enabled retention of the respective layer orientations of the multitude layers in the composite, as well, provided the third direction reinforcement to the required extent. However, on the flip side, these technologies resulted in loss of properties (quantum being unpredictable) in the planar direction, which has been a deterrent for their total exploitation.

Embracing the formative approach, enabled weaving to be chosen as a promising alternative around the world to form 3D woven composites. As is expected of, the first step has been to attempt weaving 3D fabrics on 2D looms, which has been made possible for thin 3D structures. However, increased thicknesses demanded modification in the first stage and customization by the loom manufacturer in the second stage. With these two options available, orthogonal (Noobed) and angle Interlock structures have been developed and are being researched for prospective utility in composites. While, orthogonal weaving disposed threads in three mutually perpendicular directions, the structure is actually non-interlaced, bound only at the edges. This structure is well researched due to the promising uni-directional strength properties and ease of manufacture. The angle interlock structure development requires a dobby fitment to the loom, at the least, for shedding and is being explored for ballistic applications due to the ability of the interweaves to absorb the impact force.

Another promising 3D weaving technology, less explored, is the "Dual plane shedding-based 3D weaving" wherein the yarns are disposed in mutually orthogonal directions and interlaced as well. The crux of this development lies in overcoming the hurdle of building a commercially viable machine, although few prototypes exist.

This book details about orthogonal, angle interlock and dual plane shedding-based 3D weaving technologies.

Orthogonal and angle interlock technologies are based on the 2D weaving principles, wherein the concept of the 2D weaving loom is utilized for weaving them. Based on designing, warp contributes to the X, Z directions while the weft contributes to the Y direction. Chapters. 1 and 2 discusses in detail about these technologies. The chapters are oriented towards practical weaving of these structures and the intricacies involved in it.

Chapter 3 discusses the quality control aspects that would enable qualification of 3D preforms. Preform construction parameters have been arrived at based on cloth construction principles/parameters available for 2D fabrics. To assess uniformity of construction of 3D preform, Air permeability test is proposed and a sample of this has been evaluated. Several examples of carbon preforms has been put forth as part of discussions.

Chapter 4 focusses on the concept of dual plane shedding-based three-dimensional weaving, wherein the X, Y and Z threads have independent motions and form the three- dimensional woven fabric. The chapter focusses on the possible approach for design of sub-elements such as let-off system, warp congregate sheet, dual plane shedding elements, dual plane picking elements, converging beat-up system and linear take-up device. Although there may be myriad approaches for these developments, the focus in the projected concept has been to adopt the principles of the 2D weaving loom design (which has evolved over 5000 years).

Chapter 5 focusses on the Mechanical Engineering aspects of the dual plane shedding-based 3D weaving machine. Detailed sizing calculations of sub-elements, detailed description of the construction of a prototype machine and the intricacies involved in the machine design perspective are discussed. Adaptation of the mechanical engineering principles to a customized loom design has been explained.

Epilogue provides the conclusive note by the editor and his views about the future direction that is expected from this niche area.

The book has been targeted to postgraduate students, university professors and technical personnel of the weaving industry who are well acquainted with 2D weaving, especially weave design basics and the working concepts of 2D loom. A dedicated chapter on machine design approaches for dual plane shedding-based 3D weaving is a boon for mechanical engineers and loom manufacturing industries to get the new concept of loom making as well as radically deviating approaches in the design and fabrication of the sub-elements. However, mechanical engineers will need to read and understand 2D weaving basics and the conventional loom design prior to reading this chapter.

Contents

Editor and Contributors

About the Editor

Dr. Bangalore Sridharan Sugun is currently a Senior Principal Scientist at the Center for Societal Missions and Special Technologies (CSMST), CSIR – National Aerospace Laboratories at Bengaluru. He has an interdisciplinary background with a Bachelor's & Master's degree in Textiles Technology coupled with a Ph.D. in Mechanical Engineering. He has been working in the area of polymer matrix composites research for the past two decades. Some of his early works were on Low velocity impact behavior of Polymer Matrix composites, resin transfer molding, fiber metal laminates & glass-knit composites. In the last decade, he has focused on the technology of 3D weaving. His research work revolves around evolving novel concepts for indigenous 3D weaving machinery, Near-net preforming approaches and developing creative weave designs for engineering the preforms for use in composites. His recent endeavors comprises of developing a prototype machine for dual plane shedding based 3D weaving wherein basic concepts of the technology have been practically realised, development of an automated customied take-up-at-will 24 shaft dobby loom, utilizing pleat weaving concept for weaving T-stiffeners, development of hybrid carbon-Kevlar preform for fuselage wing joint of Micro air vehicle and weaving complex geometrical shapes using 3D noobed and angle interlock weave architectures ex., gas turbine engine nozzle flap. He was a Raman research fellow at the University of Manchester, UK in 2015, where he worked on addressing fiber lay-up schemes for rotational symmetric components using robotics approach and also worked towards the development of specific Profiles for applications in composites using three dimensional braiding. He has several peer reviewed publications, conference papers & has delivered several invited talks. With participation from a private industry within the country, he is currently developing a commercially viable, production version multi weft insertion 3D weaving loom for use in textile industries. He is currently associated with the Quality Assurance and Airworthiness activities of NAL in addition to his research works at CSMST.

Contributors

Shashikant Choudhary Centre for Societal Missions and Special Technologies (CSMST), CSIR-National Aerospace Laboratory, Bangalore, India

G. N. Dayananda Centre for Societal Missions and Special Technologies (CSMST), CSIR-National Aerospace Laboratory, Bangalore, India

J. Ramaswamy Setty Centre for Societal Missions and Special Technologies (CSMST), CSIR-National Aerospace Laboratory, Bangalore, India

Prasad Shimpi CSMST, CSIR-NAL (formerly), Bengaluru, India

B. S. Sugun Centre for Societal Missions and Special Technologies (CSMST), CSIR-National Aerospace Laboratory (NAL), Bangalore, India

Nomenclature

3D dual Plane shedding based 3D weaving	Type of three-dimensional (3D) weaving
Angle interlock weaving	Type of Three-dimensional (3D) weaving
Beat-up	Pushing the last inserted thread to the cloth/preform formation point (fell)
Binder threads	Locking threads used in orthogonal and angle interlock weaving
Binders	Traversing threads used to bind several layers together
Creel	Let-off device with individual warp package mountings
Dobby/Jacquard	Types of shedding device used in the Textile industry
EPI	Ends per Inch
Fell	Cloth/preform formation point
Fibre Volume	Volumetric content of fibre in the total three dimensional volumetric space of the preform
Glass,carbon, Kevlar	Technical textiles material used for weaving
Heald shaft	Mechanical device fitted to the weaving loom and is used to lift or lower warp threads
Heald eye	Part of heald wire through which the individual/multitude of threads (as per design) passes
Layer to Layer Angle interlock	Type of Angle interlock weave
Let-off	Incremental release of warp yarns for weaving
Links and fittings	Critical joints in aircraft
Loom	Weaving device

Needle picking	Weft insertion using needle like device
Noobing	Type of Orthogonal weaving (Non-interlacing Orienting Orthogonal and Binding)
Origami	Folding approach
Orthogonal Weaving	Type of Three-Dimensional (3D) weaving
Picking	Insertion of transverse threads using suitable device such as shuttle or rapier
Pleat weaving	Folded form of weaving so as to create a riser on the fabric as in terry towel or velvet.
PPI	Picks per Inch
Preforms	Combination of 2D and 3D reinforcements to arrive at the desired shape
Pseudo sheds	Sheds created between warp sheets by suitable arrangement without the up and down movement of the threads (as would be required during regular weaving)
Rapier	Weft insertion device from package
Reed count	Spacing of the reed wires
Reed	Comb like device used to push the last inserted weft as well as hold the warp yarns parallel to each other
Shed Geometry	Depicting the arrangement of warp from the shedding zone until the fell
Shedding	Separation of longitudinal threads to create opening for Weft insertion
Stuffer threads	Non-interlaced straight threads
Stuffers	Non-interlaced threads used to reinforce the structure
Take-up	Incremental rolling of formed cloth
Tex	Means of designating yarn count.1 Tex implies 1000 mtr length of yarn weighs 1gm
Through thickness angle interlock	Type of Angle interlock weave
Tow	Bundle of untwisted filaments
Warp beam	Let-off device having several warp threads arranged parallel in sheet form
Warp	Longitudinal set of threads in cloth

Weave architecture	Drawing depicting the yarn paths in the cloth or preform
Weave design	Information required by the weaver to weave the preform/cloth
Weft	Transverse set of threads in Cloth

List of Figures

Chapter 1

Chapter 2

Chapter 3

Chapter 4

Chapter 5

List of Tables

Chapter 1
Orthogonal Weaving

B. S. Sugun and Prasad Shimpi

1 Introduction

Orthogonal weaving is the process of manufacturing 3D textile structure by inserting X, Y and Z yarns in three mutually perpendicular planes. The fundamental idea behind the process, is, to orient all the yarns as straight as possible. 3D orthogonal preforms and composites have been the subject of intensive research for various reasons such as ease of manufacture, minimum crimp, versatility in obtaining different types of preforms, utilization of existing 2D weaving machinery with minimum modifications, simplistic approach of adopting 2D weaving-based design methods, etc., The most popular of the orthogonal weaving technologies is Noobing. It is a type of orthogonal weave structure and stands for Non-interlacing Orienting Orthogonal and Binding [1], a type of 3D woven non-crimp fabric. There also exists, the possibility for the Z yarn to be at an inclined angle instead of being orthogonal.

2 Weave Structure Details

The cross section diagram of orthogonal woven structure is as shown in the Fig. 1. It consists of parallel laid X threads stacked one above the other interleaved by Y threads layer wise. X and Y threads are not interlaced but are held together in a bundle by Z yarns, which traverse through the thickness (I and II). The binding of the preform is only at the edges in each of the X, Y and Z directions. For relatively

B. S. Sugun (✉)
Centre for Societal Missions and Special Technologies (CSMST), CSIR-National Aerospace Laboratory (NAL), Bangalore, India
e-mail: sugun@nal.res.in

P. Shimpi
CSMST (formerly), CSIR-NAL, Bangalore, India

B. S. Sugun (ed.), *Practical Approach to 3D Weaving*, Materials Horizons: From Nature to Nanomaterials, https://doi.org/10.1007/978-3-030-80424-4_1

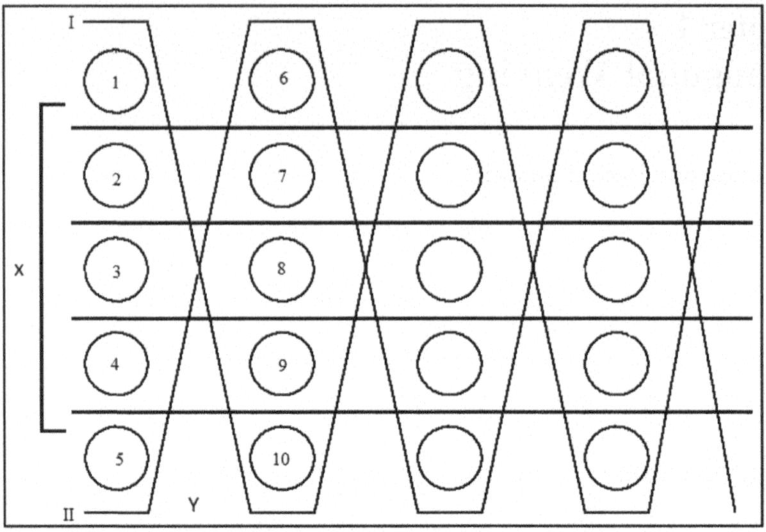

Fig. 1 Representative Orthogonal weave architecture

brittle fibres like carbon and glass, which are used in technical textiles, orthogonal woven structure is beneficial because of minimum crimp and reduced damage to the yarn.

For an orthogonal woven fabric, the primary requirement is a multi-layered warp arranged either in a grid or layered form. The number of layers depends on the required thickness and fibre volume content.

Orthogonal weaving also provides flexibility in displacement of the Z threads (Binder threads) over two or more surface picks as shown in Fig. 2b. It has the advantage of increasing fibre contributions to the X direction based on the number of surface picks it encapsulates. However, the architecture design, calculations and its weaving is complex as compared to conventional orthogonal structure. Figure 2a shows the Noobed architecture, wherein the Z yarns float over 2 successive picks.

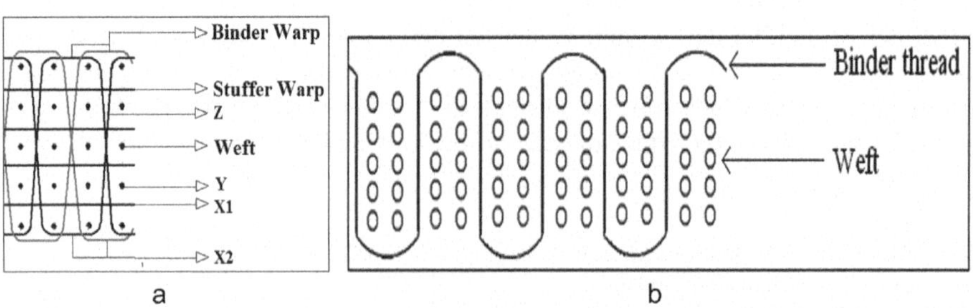

Fig. 2 A Modified orthogonal weave architecture **b** Detailed view of orthogonal structure

3 Methods of Manufacturing

Noobed structures can be manufactured using various weaving methods, which will be discussed comprehensively in this section. The simplest approach among all, is by using a perforated plate to arrange the warp threads in an array. With further increase in the thread density and complexity of preform structure, shuttle loom and its customized versions can be used. The plate weaving method and use of weaving loom for weaving NOOBED preforms is discussed here.

(a) *Plate weaving method:*

The plate weaving method has evolved from the simple concept of stacking several layers of threads in an array. These layers, are then bound together using threads in Y and Z direction, thus locking them together throughout the length and breadth of the structure.

The plate weaving method enables the development of the 3D preform in two modes viz., preforms having defined dimensions in X, Y and Z directions and preforms having undefined dimensions in X direction coupled with defined dimensions in Y and Z directions. In the former case, based on the dimensions required, the set-up is prepared and the preform developed. In the latter case, like a weaving loom, the X threads are fed and the sheet of 3D preform is developed based on requirement.

The set-up comprises of 3 dimensionally identical perforated plates arranged parallel to each other as shown in Fig. 3. The plates are aligned by guide rods such that all holes of the plates lie in the same plane.

The Base Plate has several hollow pins mounted on it to guide the X threads. It also houses the anchor points for the guide rods. Beat-up and take-up plates are assembled parallel to the base plate on the guide rods with precision matching of the perforations. However, the perforations on the Beat-up plate is required

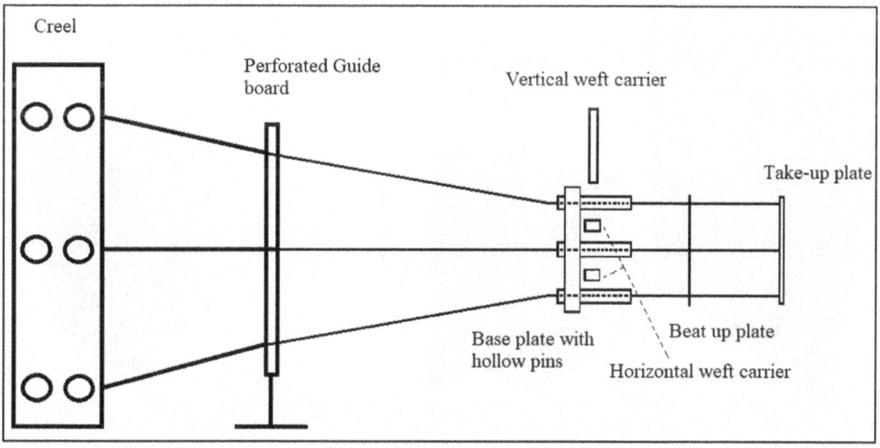

Fig. 3 Line diagram of plate weaving process

to be marginally larger than the perforations on the base plate, as this needs to slide in and out of the hollow pins for every cycle. For smooth movement of the Beat-up and take-up plate, it would be appropriate to use linear bearings. The X threads (From a creel or a miniature beam) are passed through this set-up. The major challenge lies in optimizing the hollow pin's wall thickness and spacing between the holes, as both the parameters are inversely related. With increase in the wall thickness of the hollow pin, the spacing between holes increases, which, in turn, affects the thread packing density of the final preform. The crux of the problem is to obtain as close a spacing as possible of the X threads, as this would affect the packing density. The design of the weft insertion element also plays an important role in deciding the packing density. Due to these requirements, it would be a challenge if the development calls for utilizing finer tows (such as 400 Tex, or 200 Tex) as it is usually the case in the Technical Textiles applications such as glass, carbon preforms.

The X threads when arranged in a layer wise grid pattern, as defined by the perforations, forms multiple stationary sheds between and within the layers. The hollow pins on the base plate act as guide for the weft insertion device as shown in Fig. 4b. Several variants of the weft insertion devices are available such as needles, rapiers, shuttles, etc., the design of which is governed mainly by the available space between the hollow pins. Insertion of Y and Z threads alternatively in the Y and Z planes is the preferred mode. In case needle/rapier insertion is resorted to, locking arrangement for the loop would be required. For keeping the inserted threads in their position and to prevent slipping back, a locking pin is used at selvedge, which works in synchronization with carriers and holds the inserted threads, while the carriers retrace back to their original position. A simpler approach for Y and Z thread insertion without locking

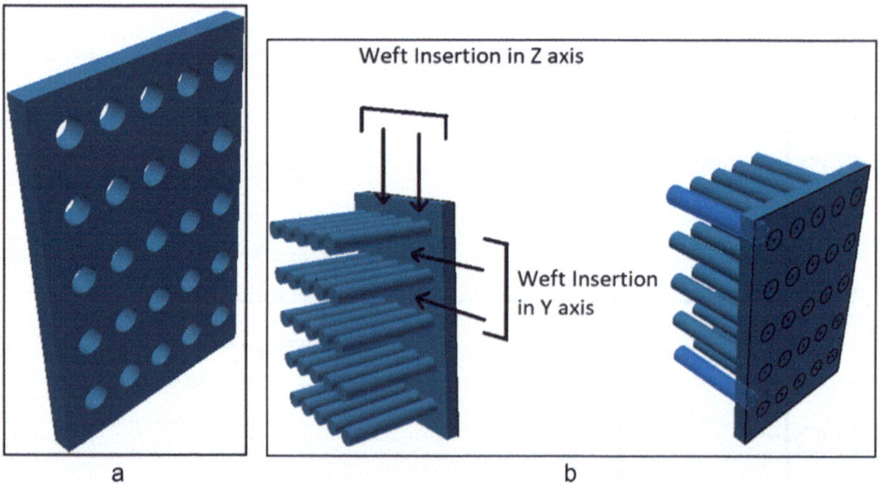

Fig. 4 Plate weaving components **a** Base plate **b** Base plate with hollow pin assembly

could be the use of specially designed shuttle of the required thickness. The self-locking selvedge of shuttle eliminates the use of additional locking mechanisms at edges. Motion of the Y, Z thread carriers can be automated by using pneumatic or mechanical system.

The beat-up plate pushes the Y, Z threads to the preform fell, similar to the conventional reed. For continuous length 3D preforms, the beat-up plate moves back and forth till the fell, similar to the conventional reed beat-up. For fixed length 3D preforms, the beat-up plate continuously recedes from the original fell point. Take-up plate can be designed to be stationery for manufacturing preforms with definite dimensions or it can slide on the guide rods for preforms with continuous length. However, for large length of preforms, conveyor-based take-up would be practical after the initial take up by the plate.

The advantage of plate weaving method is that it is simpler in construction and easier to design for specific shapes such as rectangular cross section. However, specific shape weaving like 'T' or 'I' cross section using the plate concept will predominantly be a manual process, since, locking of the Y and Z threads will not be amenable for mechanization. The other approach usually resorted to, is to manufacture blocked preforms of fixed dimensions, convert them to finished product (composites) and shape them to the required dimensions. This is not a preferred approach due to wastage of material, discontinuity in the threads, etc.,. Also, close packing density for warp cannot be achieved due to the requirement to position the pins and calls for change of set-up itself even for small changes in thread density.

(b) *Conventional Weaving loom:*

The conventional weaving loom has always been a creative machine and a platform for practical experimentation for the interested weaver/ cloth designer to explore various possibilities of weave designs and new fabrics. It is to be understood that any structure woven on a loom is consistent and repeatable with better control on thread densities and fabric quality.

The fundamental requirement to weave orthogonal 3D fabrics on the weaving loom is assigning the directions X, Y and Z to the threads. 3D preform, as is the case can be woven by assigning Z direction to either warp or weft. Henceforth, if Z direction is assigned to warp then it's termed as 'warp way weaving' and if it's assigned to weft then it is termed 'weft way weaving'. During warp way weaving, the X and Z threads come from the creel or beam and the Y threads would be the weft. In weft way weaving, only the X threads come from creel or beam while the Y and Z threads would be the weft. In some special cases discussed further in this chapter, the X direction is assigned to weft, whereas Y and Z direction to warp, respectively.

Consequent to assigning the X, Y and Z directions to the threads, the next task would be to arrive at the basic specifications of the required preform. The basic specifications required would be the dimensions of the preform (Length, Breadth and Thickness) and fibre volume contents of X, Y and Z threads. Based on these inputs, it would be required to arrive at the weaving approach (warp way or weft way), weave cross section, weave design, thread densities (EPI, PPI, No of layers) and Yarn count.

Warp way weaving requires dedicated heald shafts for the binder threads, thus reducing the number of heald shafts available for stuffer yarns. However, in weft way weaving, all the heald shafts on the loom can be assigned for stuffer warp. This would, therefore, result in the possibility of obtaining higher thickness. Dimensional accuracy would be near perfect in warp way weaving, as the binder threads have their allotted space in-between the stuffer threads. In few cases, customized reeds as shown in Fig. 5, (wherein, the reed count would be different for stuffer and binder threads) have also been used, which ensures that the obtained preform width closely matches the set preform width. In weft way approach, there is a tendency for the preform to be wider after doffing from the loom, since no dedicated space would be allotted for the binder threads in-between the stuffer threads. However, due to this aspect, weft way woven 3D preforms have the tendency to achieve higher preform density, since the binder threads would occupy the least space required during weaving in contrast to warp way weaving, as they are constrained by the reed dents. The take-up system plays an important role in achieving the preform density for both types of weaving.

Weave cross section is the primary task prior to generating the weave design, mainly because the spatial arrangement of threads will be more complex as compared to equivalent threads of 2D fabrics. In fact, based on the weave cross section, repeat size, numbering of yarns, travel path of yarns, obtainable thickness of the preform can be deciphered.

Based on the weave cross section, weave design comprising of the drawing-in details, denting order and lifting order shall be prepared for weaving. The thread densities in 3D fabrics will account for the number of layers, as well, in addition to the EPI, PPI of the surface layer. The packing density or X, Y and Z content required, decides the yarn count that can be used and the thread densities. Additionally, the specialty of 3D preforms, is that, the weave architecture also has a role to play in deciding the X, Y and Z content.

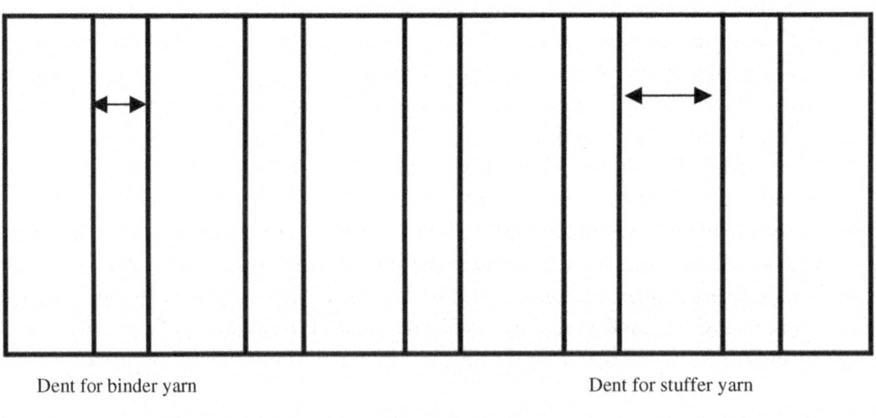

Fig. 5 Customized reed design for orthogonal weaving

Another important aspect to be considered before weaving is the denting order. To avoid congestion in the reed dent, the binder and stuffer yarns are drawn through separate dents. Binder yarns travel throughout thickness and need minimum friction in the dent whereas stuffer yarns are relatively stationery in the reed as they don't cross other yarns. Therefore, stuffer yarns (required to build the preform thickness) are threaded in seperate dent and binder yarns in separate dent. To distribute binder and stuffer yarn equally over width of preform, stuffer and binder yarn dents are kept adjacent. It is possible to weave orthogonal fabrics having stuffer and binder yarns in single dent. This will increase the preform packing density, but it is susceptible to yarn damage and increases the difficulty of weaving since stuffer thread positions gets disturbed during each cycle of weaving due to the crossing requirements of the binder threads. This damage is more pronounced in the case of technical textiles weaving such as glass, carbon, due to the brittle nature of these fibres.

4 Understanding the Constraints of 2D Weaving Looms

Prior to addressing the weaving loom requirements, it is imperative to review the constraints of 2D weaving looms (if used as it is) in weaving 3D fabrics. The main requirement of the shed forming system is to ensure sufficient shed depth for the weft insertion system to pass through. However, with increase in the number of layers for any specific weaving machine, the shed depth narrows down, and therefore, the passage of the weft insertion system gets increasingly difficult, thus constraining the number of warp layers (viz., thickness).

Weft insertion systems are designed to pass through the shed with very narrow clearance between the top and bottom shed, typically few mm. With increasing thickness of the top or bottom shed, there is every chance for the weft insertion system to get entangled with the warp. To some extent, the tension of the warp threads can counter this. However, as the number of warp layers increase (with increasing thickness of the preform), narrowing of the shed becomes inevitable, thus constraining the weft insertion system from passing through the shed. The other issue that needs to be countered by the weft insertion system is the travel distance above the race board. In conventional shuttle looms this distance is usually fixed. The weaving machines that do not have a race board will have a pre-defined path for the weft. As thickness in the bottom shed increases, it becomes a hindrance for the weft insertion system to pass through, as no weaving loom has the provision for movement of the weft insertion system in the vertical plane. This should not be confused with pirn/shuttle changing looms, as their agenda is different as there is no change in top or bottom warp sheet thickness, unlike the above case. While conventional weaving looms (shuttle based) can traverse from layer to layer, binding the selvedge, the modern shuttle-less looms will have the additional obstacle of locking the selvedge threads for each layer.

The fell of the cloth in 3D weaving is a line contact unlike 2D weaving where it's a point contact. Due to inclination of reed w.r.t vertical axis, contact point and beat-up force changes for picks inserted in various layers. As a result of this, uniformity of compaction will not be there. To keep the fell position at reed centre, either the sley height has to be reduced or there has to be a provision to elevate fell of cloth to the required height. Usually, the fell to reed distance is low. This leads to an increase in the angle and steepness of shed geometry.

Finally, the roller-based conventional take-up system will work only for thin 3D fabrics (up to 2 mm thickness). However, with increase in the thickness, the bending radius of warp close to the take-up roller and the warp away from it, will differ and will cause preform distortion. These are some of the constraints of the 2D weaving loom to use them to weave 3D preforms. It should be understood that 2D weaving looms are utilized 'as it is' to weave thin 3D fabrics.

5 Loom Requirements

The weaving loom requirements detailed below is suitable for both orthogonal and angle interlock weaving.

Let off: Let off for 3D weaving can be either from the creel or multiple beams. Use of creel has the advantage of independent tension control of the warp yarns and the disadvantage of consuming floor space. However, this is preferable for carbon and glass weaving, given the delicacy of the yarns. If resorted to let-off from beam, multiple beam approach is necessary as it separates the stuffer and binder warps as they are of different lengths in the preform. However, the beam approach is preferable for centre open sheds as the tension uniformity is best maintained.

Shedding: Shedding for orthogonal weaving is carried out layer wise, followed by pick insertion in-between each layer. In the case of angle interlock weaving, shedding shall be as per the design. A dobby or jacquard is required as shedding device. Sequential separation of layers is a pre-requisite. An added advantage during loom based manufacturing (especially for orthogonal weaving) would be to align the heald shaft/heald eye along the line of the shed geometry (preferably top line) from the fell, as this would ensure that the stuffer warp layers of the 3D warp will automatically stagger in the dents layer wise. This, eventually, will minimize the friction and damage to the warp yarns and aids in shed formation. An important point to note, is that, the shed opening sequence during 3D weaving should be from bottom layer to top layer in the case of top open shed and vice versa for bottom open shed. Considering the Top open shed as an example, wherein, the top shed is the loom home position (shed closed position), shed opening requires movement of the layers to the bottom shed position creating additional drag and tension on the warp sheet The drag and tension variation would be least if the above mentioned sequence is followed. If the shed opening is reversed (i.e., top to bottom) the drag and tension variation would be higher. This would exactly be opposite for bottom closed sheds. Such issues do not exist in centre open sheds. Summarizing, while centre closed sheds would be ideal, weaving with top or bottom closed shed is also possible and sometimes advantageous for orthogonal weaving for dropping/ lifting the stuffer threads sequentially. This, however, does not apply for angle interlock weaving.

Denting: Although the stuffer and binder threads can be dented together, it is preferable that separate dents may be used. This is to prevent damage to the stuffer threads by the criss-crossing of the binder threads in the healds. Stuffer and binder yarns are dented as shown in Fig. 7. To take into account the large number of stuffers as compared to the binders, a different type of reed design can be adopted as shown in Fig. 5, where the spacing varies for the stuffer and binder yarns.

Picking and Beat up: The preferred picking method would be using shuttles, as these form self-locking selvedges. Picking happens layer wise. There is no change in the beat-up approach. However, for thick 3D fabrics, multiple beat-up possibilities would be beneficial to achieve the required compaction. Of course, multiple beating is not usually resorted to, as this would contribute towards damage of the yarns.

Take up: Take-up for 3D weaving is required to be "At will" and is considered as one of the primary requirements. In the conventional take-up system, fabric is pulled pick by pick to maintain the set pick density as well as the warp tension. If this approach is applied for 3D orthogonal weaving, it would result in staggering and offsetting of the picks in vertical plane, thus reducing the intended thickness. Hence, "At will" take up is of paramount importance for 3D weaving and the loom is required to be adapted to meet this requirement. As a result of this, the take up at will system remains stationary for several pick cycles, such that only shedding, picking and beat-up action takes place. This ensures that the picks are perfectly stacked one above the other, as per the defined number of layers. To meet this criteria, the take-up system should be electronically controlled and equipped with a software system to facilitate defining the required number of picks and the amount of take up required. This results in a near-perfect 3D orthogonal woven fabric.

Customized 3D orthogonal weaving approach: To combine the favourable features of the plate concept (such as simultaneous weft insertion) and loom (preform packing density requirements), a customized machine for weaving only 3D orthogonal fabric can be developed. Some attempts are already carried out in this direction. The manner in which customized loom successfully combines multiple weft insertion with required thread density is explained further.

Threads are preferably taken from creel, but weaver's beam can also be used. The permanent open shed is achieved by arranging stuffer warp in various ways as follows:

(i) Passing stuffer threads through specialized heald wires having multiple eyelets.
(ii) Using soldered reed with multiple anchor points.
(iii) Staggered stationary Heald shafts in vertical plane.
(iv) Using vertically staggered smooth rods as shown in Fig. 6.

All the methods above ensure that the stuffer yarns are always stationary in their respective planes. The binder yarns are passed through 2 to 4 shafts (depending on the weave architecture). It is a primary requirement to position these shafts ahead of the stationary devices to enable crossing of the binder yarns. For efficient weft insertion, specialized needles are stacked vertically corresponding to number of sheds and shed height. These needles insert multiple weft simultaneously in all the sheds,

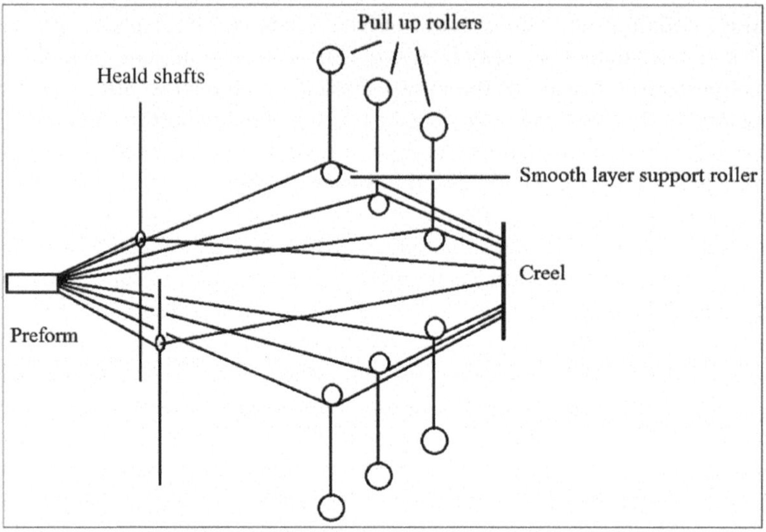

Fig. 6 Illustration of vertically staggered smooth rods

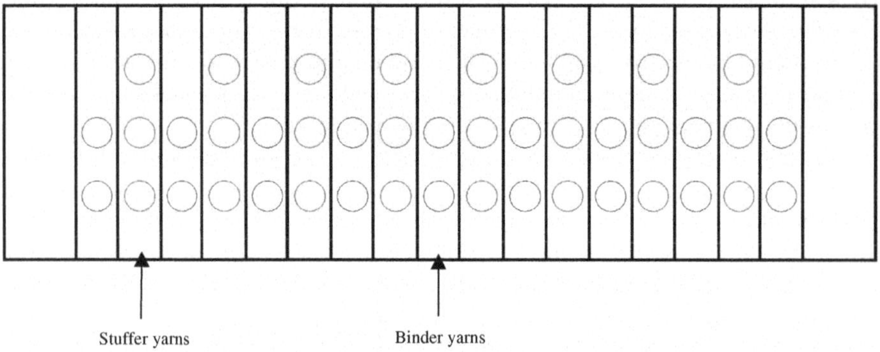

Fig. 7 Distribution of stuffer and binder yarns for orthogonal weaving

thus increasing the production and efficiency. Beat up is performed by modified conventional reed. Such reeds have large height to accommodate the permanent open shed which otherwise cannot be accommodated in conventional reeds. Due to the inherent structure, the 3D preform, being thicker, requires simultaneous beat-up of all the layers to achieve the required compaction. Hence, linear beat-up is preferable. This also takes care of the large dimensions of the reed, as arc movement is not feasible. The Beat-up distance is large due to high convergence of threads from heald shaft to beat-up point.

6 Shape Weaving Methods

Apart from imparting thickness to the fabric, objective of 3D technologies is also to manufacture preforms of required shapes. Discussions about various techniques such as pleat weaving, weft integration and origami by Orthogonal weaving to achieve 3D shapes is explained. Examples of T stiffener, integral taper, out of plane projections are also discussed. The relatively straight orientation of threads and simple structure makes Orthogonal fabric structure preferable for weaving preforms with definite shape. All the shaped preforms woven on 2D loom are in folded form, which, when opened gives the final resultant shape. A generic flow diagram as shown in Fig. 8 can

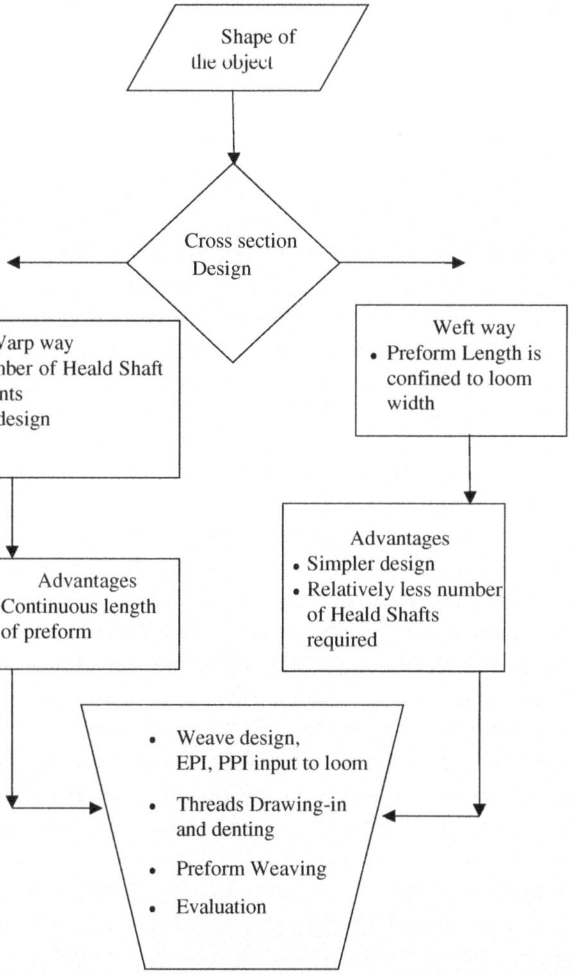

Fig. 8 Flow diagram of shaped preform design

be helpful in designing shaped preforms. With stacking of warp and weft, one above other in different ways, some guidelines are provided below, which can be helpful to design the desired shaped preform.

1. Binder and stuffer yarn should be dented in their respective places, irrespective of weave design.
2. Warp way surface integration: For shapes like "T" or "I" structure, where it is required to merge two or more surfaces in warp way; Initially, the stuffer, binder and weft yarn should be separate for individual surface. Later, at the merging point, the weave design should be changed so that stuffer merges with stuffer yarns and binder merges with binder yarn. After merging, the thickness increases provided weave architecture remains constant. The thickness difference between pre-merge and post- merge surfaces is reduced by adjusting the warp-weft ratio or architecture.
3. Weft way surface integration: For merging two or more surfaces weft way, on either side of merging point, binder and stuffer yarns must be drawn/coming from separate heald shafts. Weft remains common for both surface and thus integrates them. Thickness remains unchanged.

(a) *Pleat Weaving*

In 2D fabrics like towel and velvets, concept of pleat weaving is applied to project short strands of yarns out of the fabric plane. This is achieved by imparting variable tension to predetermined group of threads. Similar concept has been extended to orthogonal 3D weaving to manufacture basic shapes like "T" and "π". A novel 5 part weave design has been developed by Sugun and Sandeep [2] as shown in the Fig. 9. The "T" structure is divided in 2 parts, namely, the web and the flange portion. Each portion has its own individual stuffer and binder yarn sets. The weaving commences with base portion where stuffer and binder yarn sets of flange and web are combined as shown in part 1a and 1b of Fig. 9. Base weaving continues till a predetermined point, after which the web portion weaving starts. As shown in part 3a,b and c, the stuffer and binder yarns of web continue to weave as a separate fabric till the required length is achieved, while weaving of flange portion is temporarily halted. During weaving of web portion, the yarns on flange portion remain as temporary floats. These floats are adjusted by negative take up, which shifts the cloth fell from part 3b to part 4. This amounts to the exact length of woven web portion, which forms a pleat. In the last step, weaving of base portion is repeated as shown in part 5a and part 5b. Although the pleat manufactured by this method resembles 'T' shape, a gap exists in the flange portion which hampers the acceptance criteria of 'T' stiffeners. To rectify the drawback, extra stitching threads (EW) are incorporated in the flange structure while weaving. These threads extend along the straight length of the flange while it is being woven, but needs to be pulled when pleat is being formed during negative take up. To achieve uniform thickness in web and flange parts of 'T' structure, the fabric thickness of web portion is kept half of the base portion as pleat

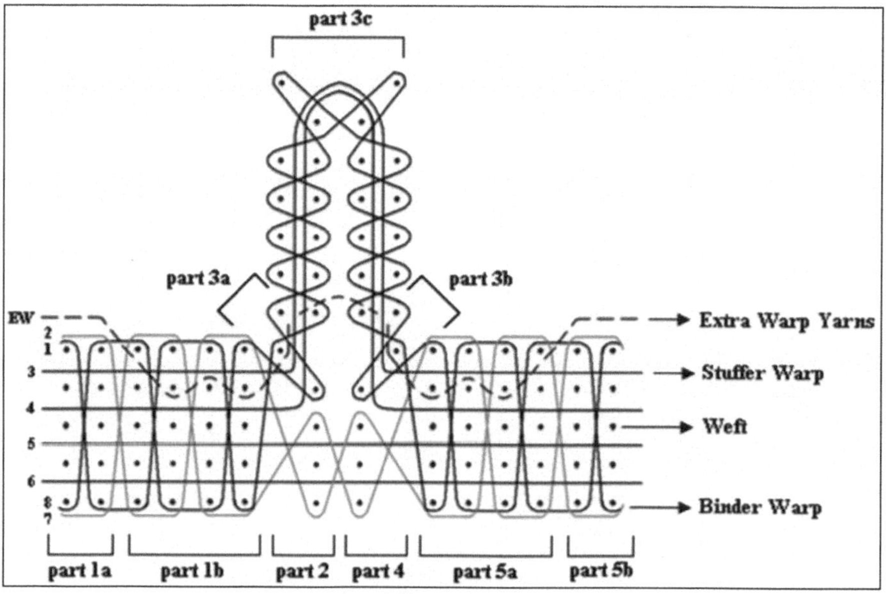

Fig. 9 Weave architecture of pleat 'T' preform

formation doubles the thickness. The length of preform is confined to the loom width and continuous length production is not possible by this method.
The weave design for all the 5 parts is different. The 2D loom on which the preform is woven must be equipped with take-up system which can reverse fabric take-up directions. Figure 10 a–e shows the actual stepwise weaving of 'T' preform on 2D weaving loom equipped with take up at will system.

(b) *Tapered Weaving*
For applications like blade of a windmill, where the component thickness is not uniform and thickness change is gradually creating a thickness gradient, special technique of tapered weaving is developed to accommodate the component requirement. Laminated composites having such thickness gradients are manufactured by decreasing/increasing the number of layers in the required zone. In 3D woven preforms, approach is similar, but techniques are different for warp and weft way weaving.

i. For obtaining a taper in weft direction (along the width), denting arrangement of warp in the reed is important as shown in the Fig. 11 Warp yarns are dented through the reed in either increasing or decreasing order of number of layers along the width. Number of binder yarns is same throughout the width, but stuffer yarns are dented in increasing or decreasing order. Weft is also inserted with gradual increment in the width.
ii. Shown in Fig. 12 is the weave architecture for obtaining taper along the length (warp direction). In this case, gradient in the thickness is obtained

Fig. 10 Stepwise weaving methodology of Pleat 'T'

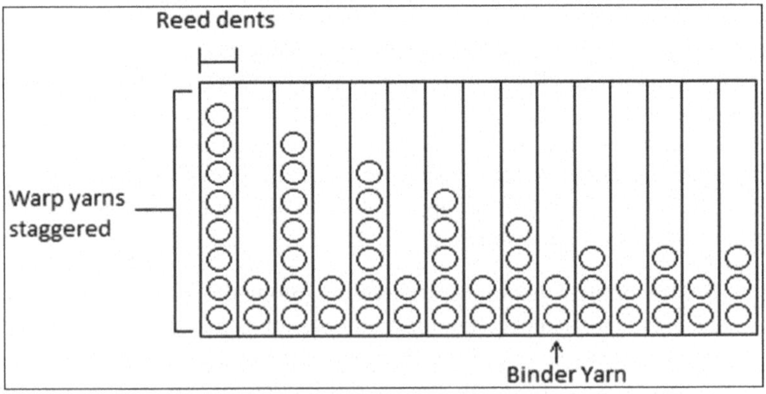

Fig. 11 Warp arrangement in reed for taper weaving in weft direction

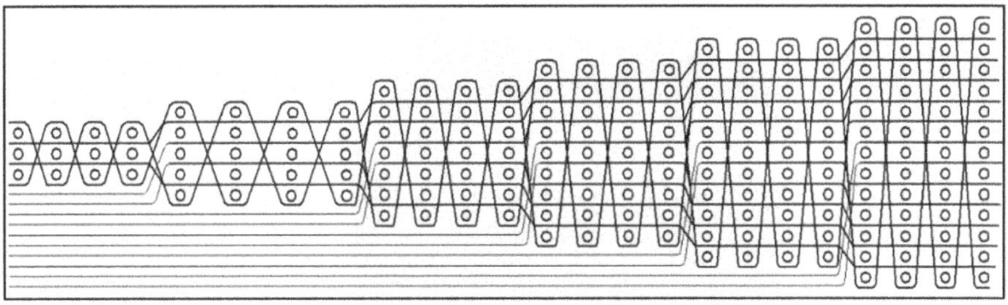

Fig. 12 Weft cross section for taper in warp direction

by increasing stuffer warp layers during weaving. As a first step, the maximum thickness decides number of stuffer warp. These stuffer warps are then woven in preform, sequentially and layer wise. The part of stuffer yarns which are not woven are left as floats during weaving and cut after the preform is woven just like the process of making a velvet fabric. The taper obtained here is in discreet steps with increase in thickness, which can be smoothened by increasing the weft density and reducing pitch of the step. After the preform is made, floats of stuffer yarn are cut and the stuffer fibres exist as small stubs of yarns rather than continuous tow/thread. During tapered weaving if every new layer of stuffer yarns are added on the surface, then these stubs tend to disintegrate when the floats are cut, deforming the fabric. To rectify this, every new layer of stuffer yarn is diverged at the core/centre portion of the preform while the relatively longer stuffer yarns remain on the surface providing a cover. Therefore, after cutting floats, smaller stubs of stuffer yarn remain at the centre without being disintegrated. Figure 13 shows the actual preform photograph.

(c) Surface integration by weft

For continuous length production of shaped preforms, integration of surface by weft is a useful technique. The stiffener component in wing structure of aircraft uses 'T' shaped element made from either metal or composite. CH Chiu and CC Cheng [3] have proposed innovative concept of using multi eye heald wire for continuous production of shapes such as 'T' on 2D loom. R Schmidt et al. [4] have patented weave design based on angle interlock fabric structure for weaving 'T' and 'π' shapes. In this section, another method to weave shaped preforms by weft integration and having orthogonal fabric structure is discussed.

Figure 14A shows the warp arrangement as on the loom (warp cross section) for weaving a single layer 'T' shaped preform. The arrangement is decided by the folding direction of the profile, density and spacing of the warp threads in the resulting woven profile. Each dot indicates all the warp threads belonging to the surface, in a single dent. Weaving of the 'T' profile is accomplished by four pick repeat cycle shown in successive figures from 14B to 14E depicting

Fig. 13 Tapered preform woven on loom

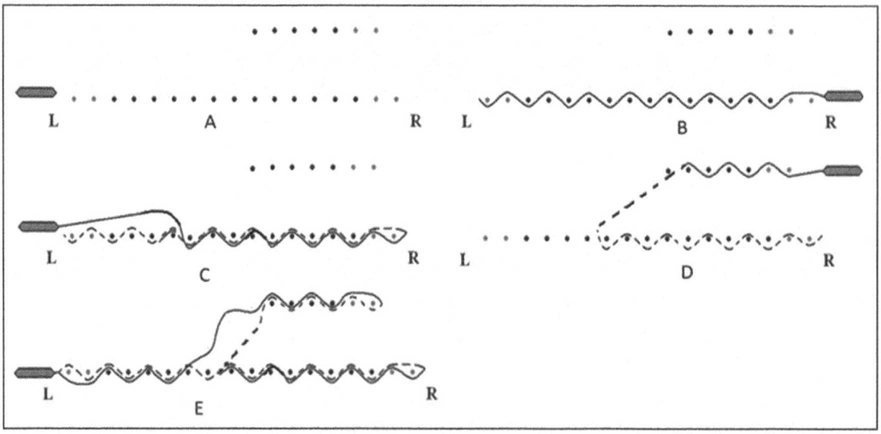

Fig. 14 Weaving methodology for weft way surface integration

first to fourth pick, respectively. In all the descriptions below, the just inserted previous pick has been shown by means of dotted lines. L to R implies that the shuttle has travelled from the left box to the right box and R to L implies vice versa, i.e., the shuttle has travelled from the right box to the left box.

1st pick (L to R)–Fig. 14B: Here, the bottom layers of the warp threads are interlaced.

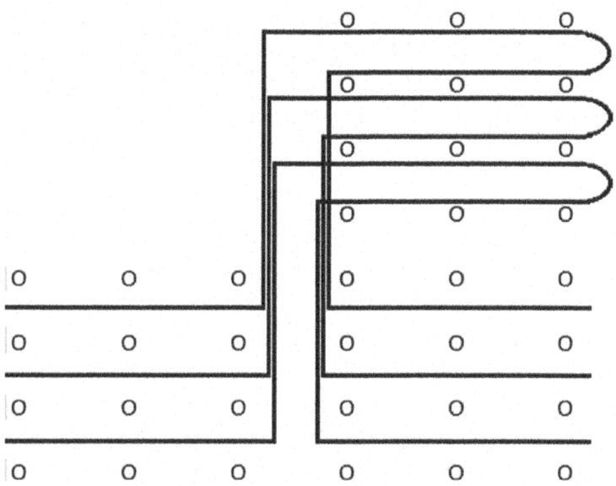

Fig. 15 Warp cross section of weft integrated 'T'

2nd pick (R to L)–Fig. 14C: The warp threads are interlaced selectively in the bottom right half of the warp threads. The left half of the warp threads are lowered down resulting in the weft floating over them.

3rd pick (L to R)–Fig. 14D: The warp threads were interlaced in the top layer continuing from the 2nd pick. The weft float of the 2nd pick was taken into the weave formation for the 3rd pick.

4th pick (R to L)–Fig. 14E: The warp threads in the top layer are interlaced and the shuttle traverses to left leaving a weft float without interlacing the warp threads in the bottom portion. The concept of weaving the above single layer based 'T' shape can be extended to orthogonal weaving as shown in Fig. 15. Here the weft does not interlace with warp, instead it is bundled together in the structure by the binder yarns.

(d) Weaving of complex shapes

The weaving techniques discussed till now is for manufacturing of simple shapes like 'T', 'L', 'π', etc. However, in majority of industrial applications, the shapes and contours are often complex and difficult to manufacture. If the component has to be made by adding layers of fabric for thickness build up, then fabrics with different architectures are preferred according to the contour and property required. In 3D orthogonal weaving technology, the previously discussed techniques can be combined to weave complex 3D shapes.

For illustration purpose, the shape as shown in Fig. 16 is taken in consideration for development. The process of weaving is on 2D loom so the 3D shape has to be converted into a 2D form which can be easily woven on the loom. This is known as Origami, where a 2D pattern is developed such that it opens to a 3D shaped object. Mapping of shape on a 2D plain sheet gives pattern as shown in Fig. 17. The dotted line shows where the fold should be done.

In context of weaving, length of the pattern can be oriented either in warp or weft direction, but weft direction orientation is simpler to design and manufacture which is discussed further in detail. The 2D pattern can be divided into smaller parts D1, D2 and D3 as shown in Fig. 18. The weave design for D1 and D2 surfaces are completely different but the fabric structure remains orthogonal. The surface D3 is

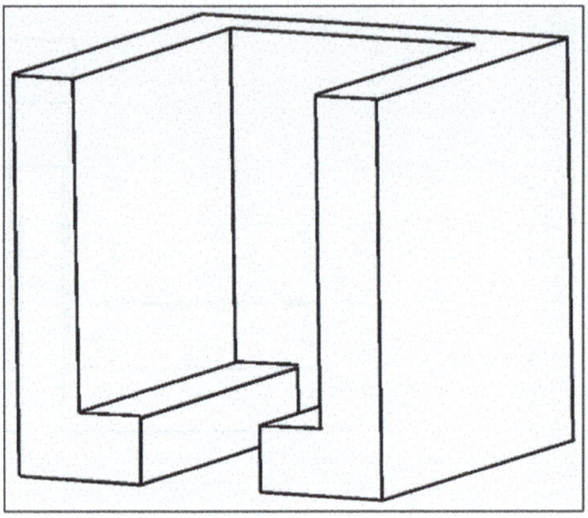

Fig. 16 Complex 3D object to be replicated

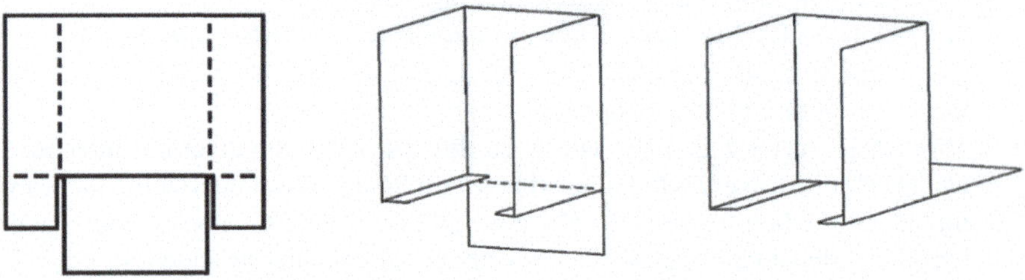

Fig. 17 2D origami pattern of the 3D object

Fig. 18 Weave planning on the loom for the 3D object

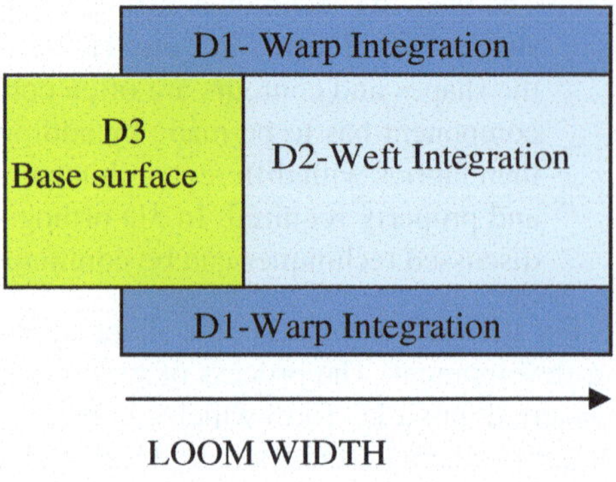

not a separate surface but the integration of D2 with the base fabric on which it is woven. The weaving commences with D1 part which is woven as separate surface, with individual binder, stuffer and weft yarns. As the weaving progresses, part D2 is woven as surface integration by weft. The stuffer and binder of D1 merge with the base fabric. In the last section of pattern, weaving of D1 is resumed and the

Fig. 19 Photograph of woven shape

stuffer-binder yarns re-emerge from the base fabric. Thus, the parts D1 and D2 are held together by warp whereas the whole pattern is merged with the base fabric by weft. During weaving, the pattern is in folded form that can be opened to the shape as shown in Fig. 19.

7 Preform Calculations

Like cloth calculations in 2D fabrics, preform calculations are required to be carried out for 3D preforms. A systematic approach to calculate the amount of X, Y and Z directional fibres in the total volume of the preform is a pre-requisite. In 2D woven cloth, the warp and weft content is determined by practically counting threads in cloth, but in 3D woven preform, counting every thread is not possible as majority of them are embedded in the structure. Therefore X, Y and Z content, fibre volume and fabric weight are calculated theoretically from EPI (ends per inch), PPI(picks per inch) and dimensions (length, thickness and width). Calculated fabric weight is verified against measured weight of the preform. However, there is no way of practically measuring the X–Y–Z content and Fibre volume content. A tolerance of 5% between calculated and measured fabric weights is an indication of the accuracy of fibre content calculations. A flow diagram of the calculation process is shown in Fig. 20.

Sample preform details are given in Table 1. The binder thread is resolved into two components 1) X_2 for X direction contribution 2) Z fibre contribution. Stuffer warp, as well as weft, are laid straight (non crimp) in their respective axis. The sample preform as shown in Fig. 21 was woven using carbon tows on a 24 shaft specially built 3D weaving loom available at CSIR–National Aerospace Laboratories.

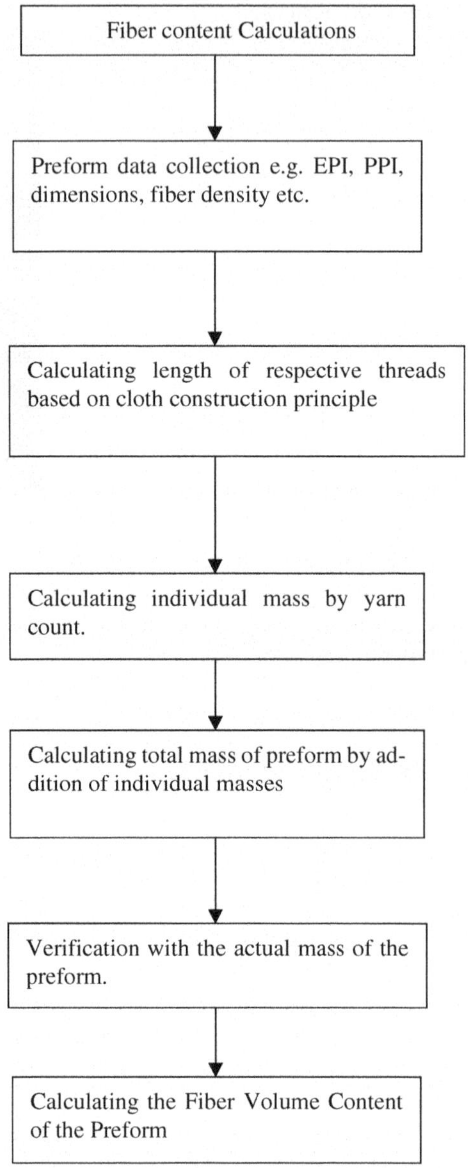

Fig. 20 Flow diagram for Fibre content calculations

Total length of stuffer warp, binder warp and weft is calculated as shown in Table 2 and corresponding value is converted into weight using yarn tex value. Fibre volume is determined using this weight and density value of the material used. EPI and PPI in 3D preform implies accounting for surface and through thickness threads together. For example, 80 EPI may imply 10 layers of warp with measured 8 ends/inch on surface. It is to be noted that partial contribution of Z threads in X

Table 1 Sample preform details

SUMMARISED SAMPLE PREFORM DATA SHEET

1. Raw Material details a. Material b. Type c. Count d. Modulus e. Density	Carbon tows Untwisted 400 Tex Intermediate – 230 GPa 1.75 g/cc	
2. Preparatory process details Warping (beam or creel) Total No of Stuffer warp Total No of binder warp	Creel 324 72	
3. Preform details	**As planned**	***Actual Obtained***
Overall dimensions (L × B × H) in Cm	23 × 23 ×0.3	*23 × 22 × 0.29*
Total ends per inch for stuffer = No of warp Layers × No. of binder dents/inch	9 × 4 = 36	*9 × 4 = 36*
Binder threads per inch = No of ends per dent × No of binder dents/ inch	8	*8*
Total picks per inch = No. of layers × surface picks per inch	10 × 6 = 60	*10 × 6 = 60*
Preform weight	90.62	*95*
Total fibre volume content	*33.97*	
X content (%)	*40.19*	
Y content (%)	*54.8*	
Z content (%)	*5.01*	

direction at the reversal points on preform surface must be accounted during fibre content calculations. 3D woven preforms consists of considerable air gaps. These air gaps determine the air permeability and density of the preform. The total mass and total volume of the preform comprises of mass and volume of fibres and air entrapped in the pores of preform, respectively. As the mass of air is negligible compared to the mass of the fibres, therefore, mass of the preform is safely assumed to be comprised only of fibres. Whereas, the volume occupied by air is significantly large, therefore, it cannot be neglected. To effectively estimate the content of air and fibre, the term Fibre Volume Fraction (Fvf) is used. It is defined as the ratio of total volume of the fibres in relation to the volume of the preform. During making of composite from the preform, air is replaced by matrix, so Fvf directly shows percentage of fibre to matrix. This is the deciding factor for properties of composites. Taking the above points in consideration, the calculation of Fvf is shown in Table 3.

The reader is requested to refer the chapter on Quality control aspects of 3D preforms and reference 3 of this chapter for additional details.

PREFORM CALCULATIONS.

Calculations (summarized) X, Y, Z contents and Fibre volume content.

Table 2 X, Y and Z content calculations

Content of warp (X)	
Content of stuffer warp (X1)	
Total number of stuffer warp (stuffer epi × width)	324
Total Length of stuffer warp in m (No. of threads × length)	72.49
Weight of stuffer warp in gm (400 tex)	29.79
Content of Binder warp (X2)	
Total number of binder warp (binder epi × width)	72
Total Length of binder warp in m (No. of threads × length)	16.55
Weight of binder warp in gm (400 tex)	6.62
Content of warp (X) in gm (Stuffer + Binder)	**36.42**
Content of weft (Y)	
Total number of wefts (PPI × length)	543.31
Total Length of weft in m (No. of weft × width)	124.15
Weight of weft in gm (400 tex)	49.66
Content of weft (Y) in gm	**49.66**
Content of Z threads	
Total number of binder warp [1]	72
Number of Z threads for one binder thread [2] (surface PPI × length)	54.33
Total length of Z threads (Thickness × [1] × [2])	11.34
Weight of Z threads in gm (400 tex)	4.54
Content of Z threads in gm	4.54

Table 3 Fibre content calculations of Preform

Total Weight of the Preform (X + Y + Z) = 90.62		
Fibre **Volume Fraction**		
Total Preform Volume in cc (length × width × thickness) **[1]**	152.41	
Total Preform Fibre Content by Volume (weight/ density) **[2]**	51.78	
Total Preform Fibre Volume Fraction ([1]/[2])	**0.34**	**33.97%**
Volume of X Fibres (warp weight/density)	20.81	
Content of Warp–X (Vol of X/[2])	**0.4**	**40.19%**
Volume of Y Fibres (weft weight/density)	28.38	
Content of Weft–Y (Vol of Y/[2])	**0.55**	**54.80%**
Volume of Z Fibres (Z weight/density)	2.59	
Content of Z thread–(Vol of Z/[2])	**0.05**	**5.01%**

Fig. 21 Photograph of Noobed Preform

8 Sample Weaving

The dobby loom is a versatile weaving loom and this versatility is utilized to weave several variants of the Noobed preform. It is to be understood that these are representative architectures based on NOOBING and myriad combinations are possible Figs. 22, 23, 24, 25, 26, 27, 28, 29, 30, shows the fabric architectures, whereas Fig. 31 shows the actual woven samples. Table 4 details the fibre content and thread density of all the 9 samples.

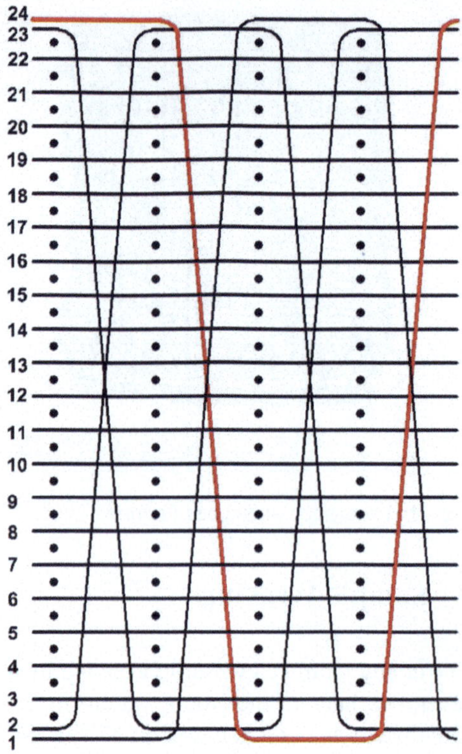

Fig. 22 Weave Architecture of Sample 1: 20 stuffer warp, 4 binder warp and pick density with 8 × 21 (ppi)–preform thickness 7.4 mm

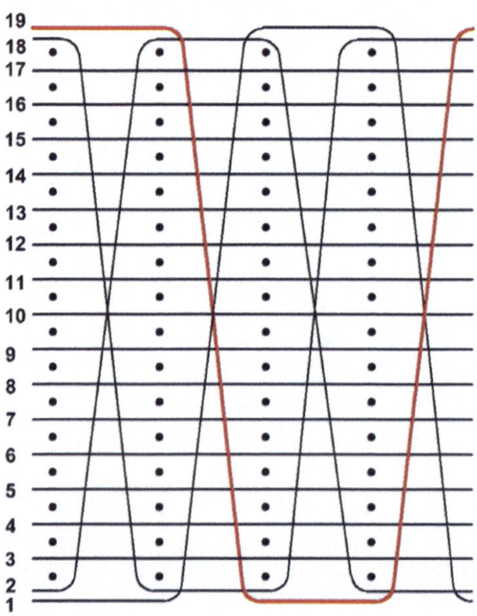

Fig. 23 Weave Architecture of Sample 2: 15 stuffer warp, 4 binder warp and pick density with 8 × 16 (ppi)–preform thickness – 5.7 mm

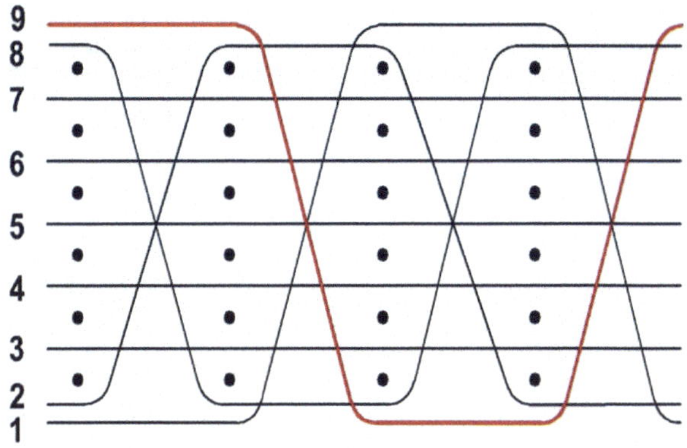

Fig. 24 Weave architecture of Sample 3: 5 stuffer warp, 4 binder warp and pick density with 8 × 6 ppi, preform thickness 2.4 mm

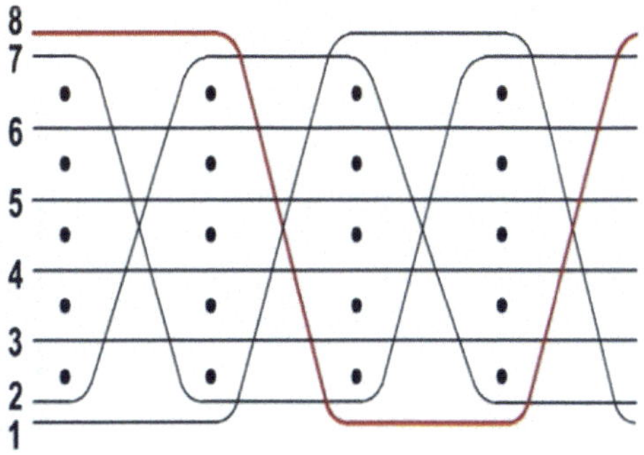

Fig. 25 Weave architecture of Sample 4: 4 stuffer warp, 4 binder warp and pick density with 8 × 5 ppi Preform thickness 2.2 mm

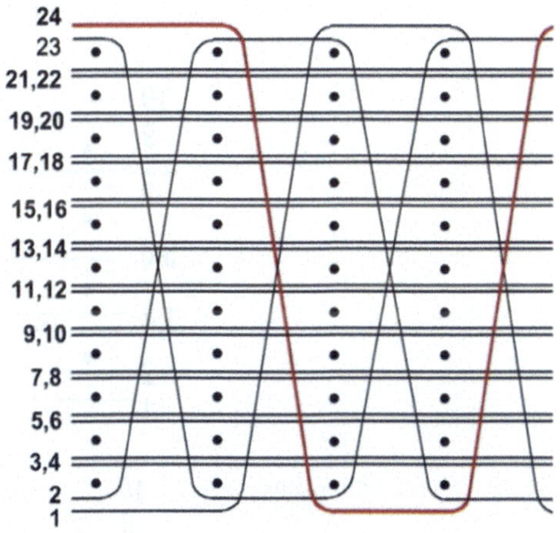

Fig. 26 Weave architecture of Sample 5–10 × 2 stuffer warp, 4 binder warp and pick density with 8 × 11 ppi preform thickness 5.3 mm

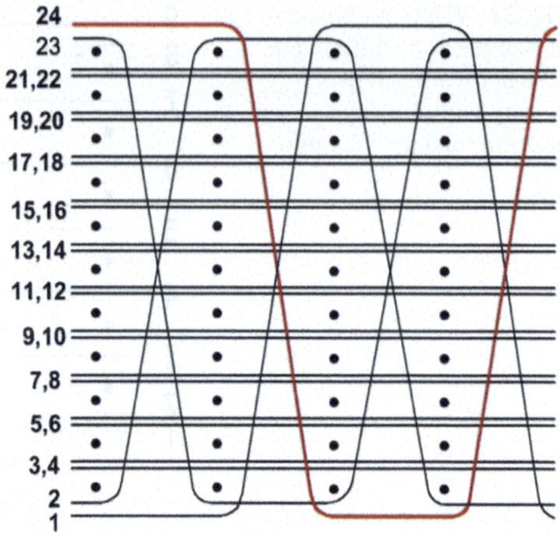

Fig. 27 Weave architecture of Sample 6–10 × 2 stuffer warp, 4 binder warp and pick density with 16 × 11 ppi

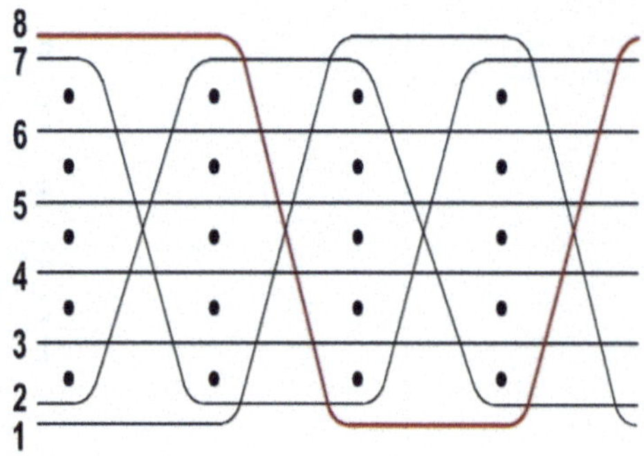

Fig. 28 Weave architecture of sample 7–4 stuffer warp, 4 binder warp and pick density with 12 × 5 ppi preform thickness 2.2 mm

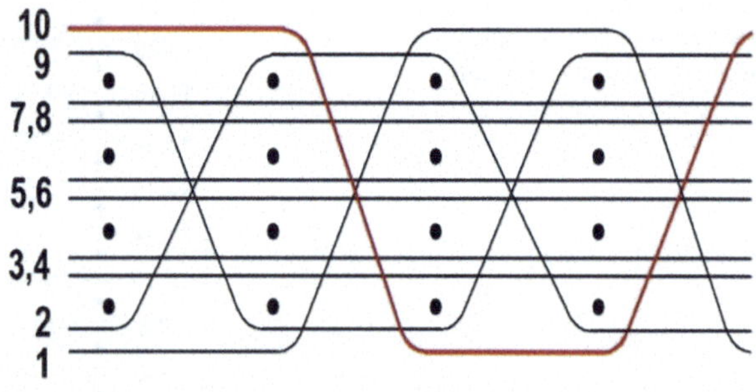

Fig. 29 Weave architecture of Sample 8: 3 × 2 stuffer warp, 4 binder warp and pick density with 8 × 4 ppi

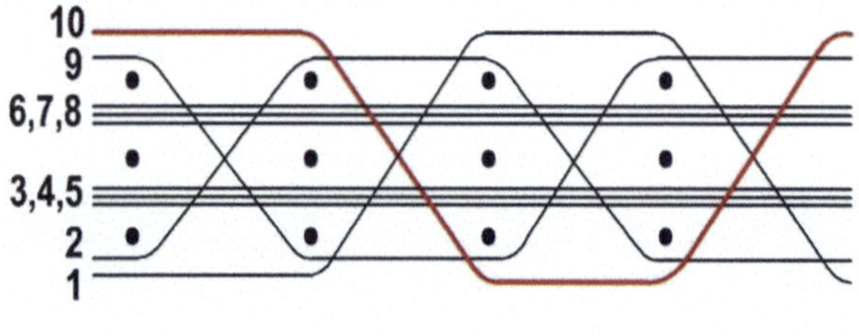

Fig. 30 Weave architecture of Sample 9: 2 × 3 stuffer warp, 4 binder warp and pick density with 8 × 3 ppi

Fig. 31 Photograph of woven samples

Table 4 Sample details for varied noobed architecture

Sample no	Thickness in mm	Stuffer warp	Binder warp	PPI	Overall fibre volume fraction (%)	X %	Y %	Z %
1	7.4	20	4	8	48.31	48.39	42.24	9.37
2	5.7	15	4	8	48.74	49.24	41.47	9.28
3	2.4	5	4	8	49.53	54.52	36.34	9.13
4	1.7	4	4	8	59.59	56.86	35.54	7.58
5	5.3	10X2	4	8	52.08	62.61	28.7	8.68
6	7.4	10X2	4	16	53.82	43	40	17
7	2.2	4	4	12	55.26	47.39	44.42	8.18
8	2.2	3X2	4	8	50.35	65.01	26.01	8.98
9	2	2X3	4	8	51.33	70.15	21.04	8.81

Acknowledgements The following students and their institutes are acknowledged for their contributions vide their project works.

1. Mr. Atharva Sanjay Shende, Internship, B Tech, Conceptual weaving of tapered 3D preform using carbon tows. Institute of Chemical Technology, Mumbai, 2018

2. Mr. Prasad Shimpi. M.Tech Dissertation work titled Conceptual Development of Roller based Pseudo sheds and biased weaving ± 45 approach for 3D Noobing. The Maharaja Sayajirao University of Baroda, 2017

3. Aravinda S, M Tech Dissertaion work, A study on modifying 3D rapier weaving machine to produce 3D Noobed structure and designing of Noobing machine. GSKSJT Institute, Visvesvaraya Technological University, 2015–16

4. Vivek G V, M Tech Dissertation, Development and evaluation of 3D orthogonal reinforcements with variation in fibre content. GSKSJT Institute, Visvesvaraya Technological University, 2014–2015

5. Bharath B N, Divyashree D and Shilpashree L. B Tech project report, Development and evaluation of 3D Noobed Preforms. GSKSJT Institute, Visvesvaraya Technological University, 2013–14

6. Shwetha singh N, M Tech Dissertation work titled Development and evaluation of 3D orthogonal T shaped preforms for use in composites. GSKSJT Institute Visvesvaraya Technological University, 2013–14.

References

1. Khokar N (2002) Noobing: a nonwoven 3D fabric-forming process explained. J Text Inst 1(93):52–74
2. Sugun BS, Sandeep DN. Integral weaving of Orthogonal 3D T Stiffeners based on pleat weaving concept. J Ind Text, https://doi.org/10.1177/1528083717702009
3. Chiu CH, Cheng CC (2003) Weaving method of 3D woven preforms for advanced composite materials. Text Res J 1(73):37–41
4. Schmidt R, Bersuch L, Benson R, Islam A. (2004) Three-dimensional weave architecture USA Patent US 6(712), 099 B2

Chapter 2
Angle Interlock Weaving

B. S. Sugun

1 Introduction

The 2D weaving technology has matured sufficiently today, for the weaver to understand and implement the variations required in a fabric. To cite an example, the thread density changes are easily carried out by changing the reed (for warp density) or pick density, prior to or during weaving, of course, depending on the weaving limits. Although the same analogy of changing the Reed (for warp density) and pick density can still be implemented even in 3D fabrics, however, due to the presence of the binder thread or the Z component, the implementation is much more complex. Most importantly, today, 3D fabrics are developed on modified 2D looms (the extent of modification varies widely based on the end use) wherein, more or less, the weaving principles do not vary much compared to 2D weaving. In this context, it would be required for the weaver to understand the possibilities of weave architecture variations, thickness possibilities and the factors affecting the above. Taking the 3D angle interlock weaving approach, as a case study, this chapter attempts to understand the thread content variations in relation to weave architecture changes for 6 K carbon tows.

Angle interlock weaving is one of the largely researched options of 3D weaving for use in composites next to Orthogonal weaving. This is because, the angle interlock preform, to a limited extent, can be developed on the conventional weaving loom with the dobby or jacquard attachment, and the advantage is that myriad weave architectures can be obtained by changing the weave designs. Preform mouldability coupled with drapeability with promising applications in the area of ballistics is a notable feature for exploration of this technology by researchers. The layer-to-layer structure enables a convenient design of the preform at curvatures, the interweaves support

B. S. Sugun (✉)
Centre for Societal Missions and Special Technologies (CSMST), CSIR-National Aerospace Laboratory, Bangalore, India
e-mail: sugun@nal.res.in

B. S. Sugun (ed.), *Practical Approach to 3D Weaving*, Materials Horizons: From Nature to Nanomaterials, https://doi.org/10.1007/978-3-030-80424-4_2

ballistics and the option for increasing the fibre content by use of stuffer threads, enables improvement of properties on the structural front. The following paragraph (not limiting) cites a few of the research activities that are ongoing in this domain.

Callus [1] et al. have carried out tensile tests on glass-reinforced polymer composites and observed crack formation at low strains within the resin-rich channels, fibre tows and around the through thickness binder yarns in the composites. The composites failure was localized and observed to have occurred by the rupture of load-bearing tows. Chen [2] has carried out a review of the mathematical modelling approaches of 3D woven fabrics. He has discussed about various software programmes and is of the opinion that they can be used to create final weave, weaving instructions and visualization of 3D weave architectures. Kadir [3] demonstrated a new method of weaving using a prototype of a multiaxis 3D weaving machine. He expressed the presence of a technical hurdle for the beat-up approach. Endruweit and Long [4] have studied the compressibility and permeability of 3D weave architectures, related to the textile architecture and compared them with equivalent 2D fabrics. They are of the opinion that through thickness permeability of 3D preforms are higher than 2D fabrics. Mahadik and Hallett [5] have studied the effect of compaction and yarn waviness on the compressive properties of 3D woven angle Interlock (AI) composites. It was observed that compaction of the fabric in the thickness direction affected the level of yarn waviness in the warp, weft and warp weaver yarns differently and therefore different trends in compression strength versus fabric compaction level were observed in the warp and weft directions. Dai et al. [6], have studied the influence of fibre architecture on the mechanical behaviour of 3D composites. The tensile properties of 3D woven composites were affected by the waviness of the load-bearing fibres. They observed that, AI composites exhibited the best performance on tensile, compressive and flexural loading as compared to orthogonal structures, but it also had the longest delamination crack among all weaves under both compression and bending loading. Patrick et al. [7] have studied the homogenization of the elastic behaviour of a layer-to-layer angle interlock composites. They have developed a new numerical tool for the mechanical study of carbon epoxy layer-to-layer angle interlock composite. Archer et al. [8] have monitored the strain in tensile test specimens of 3D composites and the embedded sensors were utilized for cure monitoring as well. Zhang et al. [9] have studied the influence of textile architecture on macroscopic response using a multi-scale model and also investigated the flexural response of a quasi-statically loaded layer-to-layer angle interlock structure. They have opined that, fibre tow kinking and breakage are considered as two major failure modes of 3D textile composites subjected to flexural loading. Gras et al. [10] have studied the macroscopic elastic parameters of a layer-to-layer interlock woven composite and have evolved a simple methodology for identifying elastic properties (or more generally stiffness) of complex materials and structures under arbitrary loadings. Chen and Yang [11] have proposed angle interlock woven preform as an alternative to plain-woven preform, without the need to cut and sew, especially at the bust portion for women's armour. Jin et al. [12] have established a microstructure model of the 3D Angle Interlock woven fabric at the fibre tow level. Both experimental and finite element approaches have been adopted and found to closely match the ballistic penetration behaviour. Nawab et al. [13] have studied the changes that occurred in 5 layer and 13 layer angle interlock fabrics when moulded at five different angles and two different bending radii. The purpose has been, to understand preform mouldability. They have observed that changes related to filament distribution and tow geometry have a direct correlation with macro-scale changes viz., angle of bending and thickness of the preform. They have concluded that mouldability-based changes have a prominent effect on the mechanical properties of the composite. Sun et al. [14] have investigated the stress distribution and deformation of one type of Layer-to-Layer angle interlock composite under three-point bending cyclic loading. They have observed that, the top and bottom layers along the thickness directions undertake the majority of the load during fatigue tests. Yang et al. [15] have investigated the fatigue behaviours of the 3D woven composite undergoing three-point bending from experimental and FEA approaches. They have found good agreement in the damage morphologies and stiffness degradation. Buchanan et al. [16] have used weaver and material data to calculate the geometrical characteristics of the 3D woven unit cell for an angle interlock-type binding configuration. They have concluded that ,to measure all the irregularities that result in an intended departure from the intended design of the weave architecture will be difficult. Walter et al. [17] have proposed a new test method for performing dynamic short beam shear tests using a momentum trapped Hopkinson pressure bar. They have studied the effect of loading rate on damage evolution and have observed a change in load–displacement response, initiation and propagation of damage characteristics. Gregory et al. [18] have studied the matrix-cracking pattern in Chemical Vapour Infiltrated composites for a wide variety of weave architectures including angle interlock structures. They have found that, maximizing fibre volume fraction and minimizing the presence of perpendicular mini-composites in critical high-stress regions leads to significant increases in component performance. Cui et al. [19] have simulated the penetration of a hemispherical steel projectile into a 3D angle interlock woven composite. They have used fibre inclination model to disassemble the 3DAWC into four inclined UD laminae. They have found that the residual velocity of the projectile and the damage morphology of the lamina are in good agreement with the experimental results and have claimed time efficiency compared with the FEA, based on the actual microstructure. Jin et al. [20] have reported the three-point bending fatigue behaviour of 3D angle interlock glass/polyester composite. They have obtained the SN curves under different bending stress levels. They have observed that progressive resin cracking, yarn transverse cracking, resin yarn interface de-bonding and yarn breakages were the main fatigue damage mechanism in 3DAWC under three-point bending loading. Marie et al. [21] have tried to evaluate the loss in mechanical properties of para-aramid warp and weft yarns from spinning stage to 3D woven preform conversion and have observed that there is a degradation of 10% for warp and about 6% for weft. Khan and Umer [22] have studied the viscoelastic compaction response and are able to predict the single-step and multi-step viscoelastic response on 3D fabrics. Zhao and Jiao et al. [23] have studied the influence of fabric architecture and weaving parameter on the thermal conductivities of 3D woven composites and have observed that angle interlock woven structures have one of the highest thermal conductivity, wherein, the fibre orientation also mattered. Sugun and Sandeep [24] have established calculations for pleat woven 3D 'T' stiffener element.

2 Weave Architecture Basics,variants and Weaving Approach

As the name suggests, angle interlock architecture is designed such that each yarn travels from its originating plane to subsequent layers at an angle, so that, all the layers are interlocked with each other and contribute to the thickness build up of the fabric. Based on the depth of interlocking, angle interlock weaving can be further categorized into two variants: Layer-to-Layer angle interlock and Through Thickness angle interlock as shown in Fig. 1a and b, respectively. In through thickness variant, each yarn traverses along the thickness from surface to surface thus contributing to X or Y and Z content simultaneously (Fig. 1b). If the angular disposition is along the warp, the contribution is towards X and Z, if along weft, the contribution is towards Y and Z. The specific angle at which each thread traverses depends on the thickness and number of layers. The calculation of angle is discussed in further sections. In the layer–layer variant, yarns traverse only to the next adjacent layer and return to its original plane. Angle interlock weave architecture requires modified 2D weaving

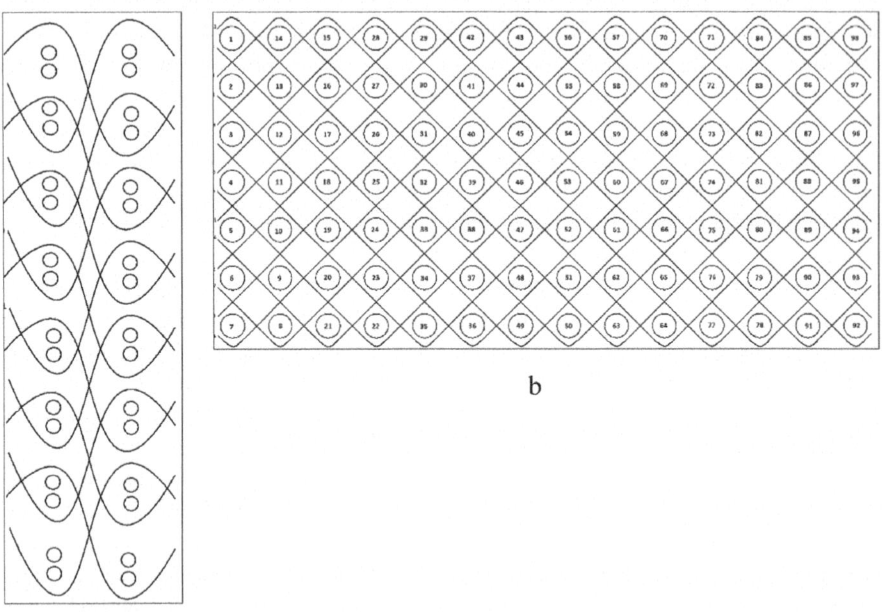

Fig. 1 Weft cross-section of **a** Layer-to-layer angle interlock, **b** Through thickness angle interlock

loom with atleast a dobby attachment for manufacturing, as opposed to orthogonal weave architecture that can be woven on different types of customised setups apart from the weaving loom. To enhance properties and fibre volume content of angle interlock 3D fabric, the weave architecture can be modified by introducing extra stuffer yarns. For illustration, 9 through thickness angle interlock weave architectures (Figs. 4, 5, 6, 7, 8, 9, 10, 11, 12) were devised and 1″ width samples were woven using a creel and multi-beam rapier weaving machine [25]. As an example, the figures represent the weft cross-section, wherein, the dots represent weft, the straight lines (where available) represent stuffer warp and the angular lines represent binder warp. Referring to Fig. 4, yarn numbers 3 and 6 are stuffer warp and the remaining are binder warp. The objective of choosing the weave architectures has been to understand the boundary of the practically achievable overall fibre volume content and X, Y and Z content. The creel is specially designed to unwind carbon tows, wherein, indirect tensioning is provided so as not to damage carbon. The reader is requested to refer to chapters 4, 5 for further details on the tensioning system for carbon. Figure 2 shows the setting up of the warp from the creel to the loom. However, the creel occupies a large space as compared to the number of warp threads mounted. To counter this problem, yarn is wound on weaver's beam that is schematically arranged to aid layer-wise separation, as shown in Fig. 3a. However usage of beams, while it reduces space occupation,the added problem during weaving, there would be the differential tension of the warp threads along the width of the beam, which would be difficult to resolve. Hence, creel would be the preferred mode, wherein, the individual warp thread tension can be adjusted based on requirement. For small widths and sample weaving, such as used here, beam weaving can be adapted. In this particular work, beam is used for few samples and creel is used for others. Possibility exists to use both beam and creel. For ex., the stuffer yarns (which just go straight into the loom and preform) can come from the beam, while the binder yarns can come from the creel. Figure 3b shows the side view of the warp beams with stainless steel thread support rollers. For carbon weaving it is required that all the surfaces it encounters is either ceramic or stainless steel. This is to minimise damage during weaving. Sample 5 (Fig. 8) and sample 9 (Fig. 12) which are the extremities with regard to X and Y content were chosen for full-width integral T weaving which is further discussed in the next section. A similar approach as detailed in the reference [24] is adopted, with the exception of the change in weave architecture from orthogonal to angle interlock architecture. The perforated plate provided in Fig. 2 is an intermediary support for the warp layers. This helps in uniform layer wise seperation and identification of threads in case of breakage or tension changes during weaving.

3 Fibre Content Calculations

In practice, there is a wide array of shapes and contours in which the preform has to be manufactured. It is necessary that the derived calculations conform with the

Fig. 2 Creel set-up with warp drawn into the loom through the perforated sheet

actual fibre volume content if the base weave of the preform is angle interlock. Due to the weave architecture of angle interlock structure, the binder thread contribution requires to be resolved into the X and Z components, where stuffer warp is incorporated, the there will be further two components in the X component viz., X_1 for stuffer part and the resolved portion of the X_2 threads for binder part. Referring to Fig. 13, the X_2 component of the binder thread is along the X direction, which on resolution based on vector principles will be X cos θ. The Z component will be X sin θ. Since, for one repeat of the weave, the binder yarn from the surface traverses to the bottom and then to the top, Fig. 13 shows the half length of the binder thread. With this backdrop, based on the vector resolution (Fig. 13), fibre content calculations are explained in further detail citing the example of 'T' woven profile shown in Fig. 14. It is to be understood that every attempt is made to give clarity to the readers regarding the calculations. However, due to the shear complexity of the calculation, the reader is required to practically work out himself to understand and appreciate the intircacies involved.

Based on the above concept, X, Y and Z content of the woven samples is calculated and tabulated in Table 1. The detailed calculations for the samples are not discussed. However, Detailed calculations are explained further in the section taking the T stiffener as an example, which is more complex compared to the samples.

Some basic trends that can be inferred from Table 1 are as follows:

Tow details: TC 33, Intermediate modulus 6K carbon
Let off: creel based
Reed count: 16s Stockport

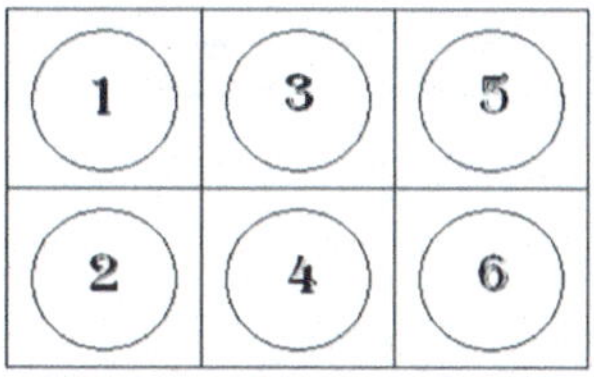

a

b

Fig. 3 **a** Warp arrangement on beam (beam 1, 2 face the loom), **b** Beam stand with warp prior to drawing-in

Samples

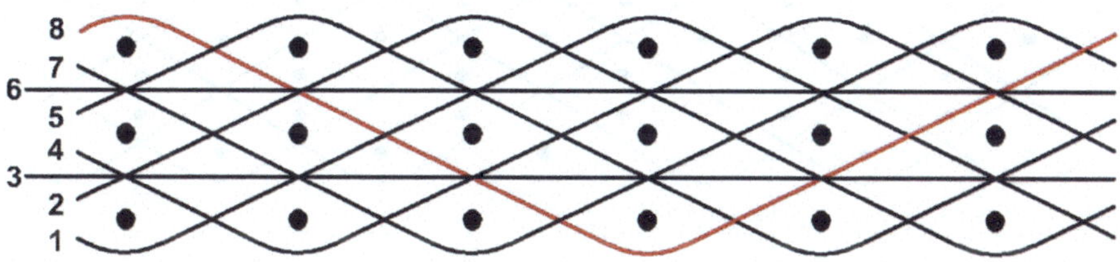

Fig. 4 Weave architecture of sample 1 (2 stuffer warp layers, 6 binder warp and pick density with 8 picks per inch)

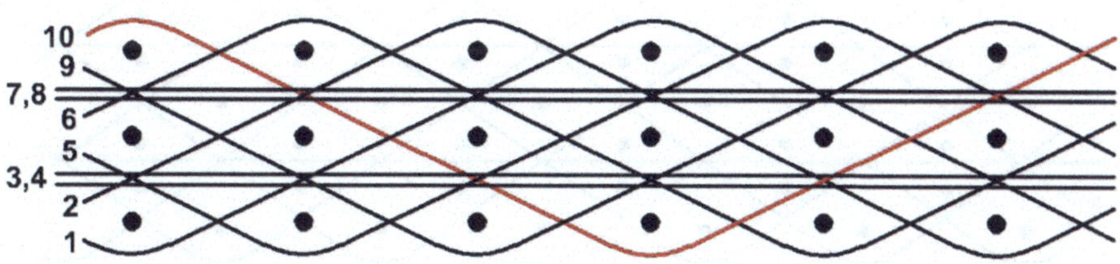

Fig. 5 Weave architecture of sample 2 (2 sets of stuffer warp in two layers, 6 binder warp and pick density with 8 picks per inch)

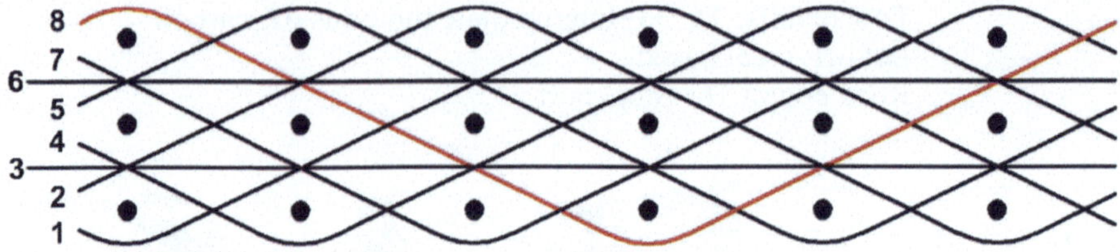

Fig. 6 Weave architecture of sample 3 (two stuffer warp layers, 6 binder warp and pick density with 12 picks per inch)

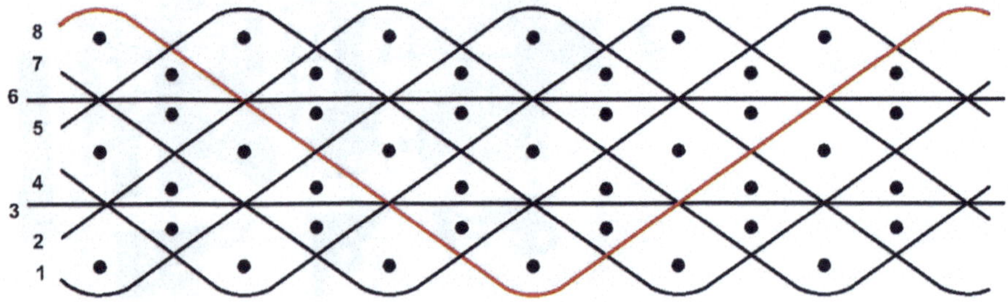

Fig. 7 Weave architecture of sample 4 (2 stuffer warp, 6 binder warp and pick density with 8 picks per inch.)

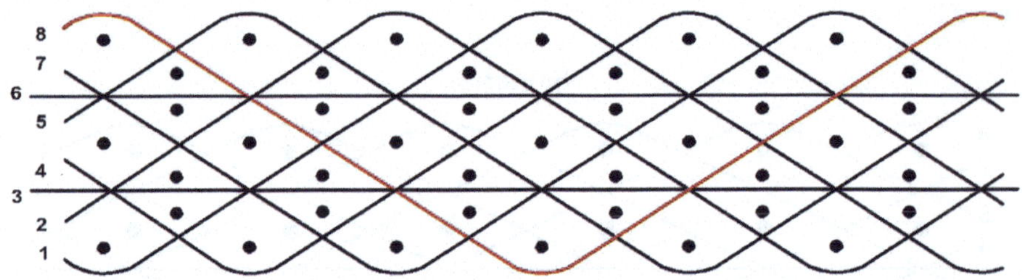

Fig. 8 Weave architecture of sample 5 (2 stuffer warp layers, 6 binder warp and pick density with 12 picks per inch)

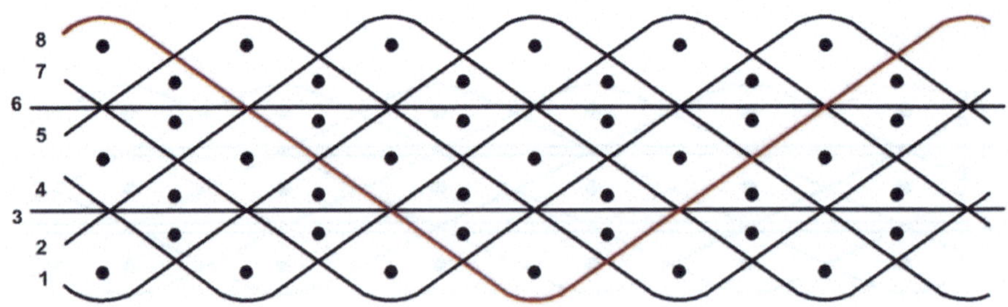

Fig. 9 Weave architecture of sample 6 (2 stuffer warp layers, 6 binder warp and pick density with 16 picks per inch)

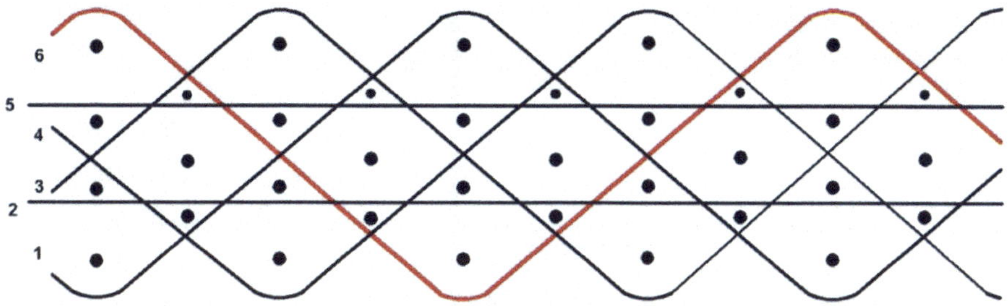

Fig. 10 Weave architecture of Sample 7 (2 stuffer warp layers, 4 binder warp and pick density with 16 picks per inch)

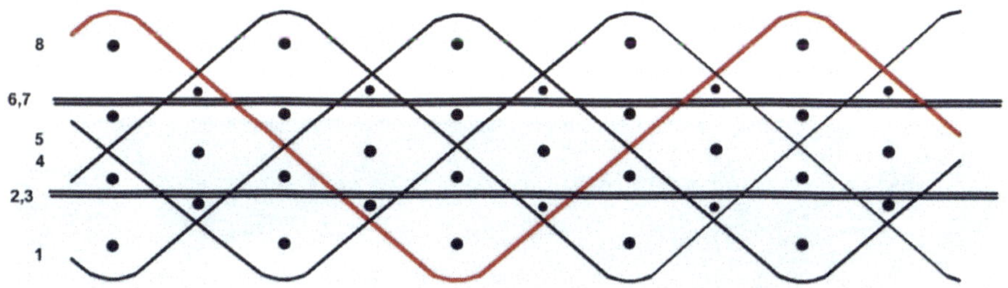

Fig. 11 Weave architecture of sample 8 (two sets of stuffer warp in two layers, 4 binder warp and pick density with 16 picks per inch)

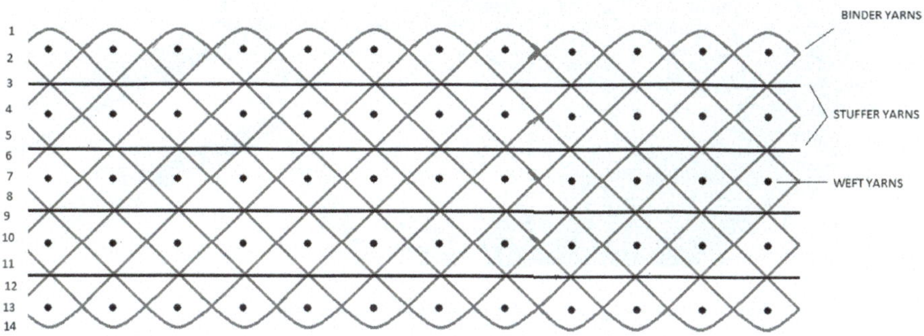

Fig. 12 Weave architecture of sample 9 (4 stuffer warp, 10 binder warp, 12 PPI)

1. Increasing Stuffer warp content (sample 2, sample 4) increases X and Z content and reduces the Y content, for an equivalent overall fibre volume content.
2. Samples 4, 5 and 6 have the same weave architecture, but increasing weft density. This has resulted in a gradual increase of fibre volume, decrease in X content and increase in Y and Z content.
3. In sample 3 and 5, weave architecture change haschanged the Z content when all other parameters are kept constant (Table 1). Development of such structures,

Fig. 13 Vector resolution of threads in angle interlock structure

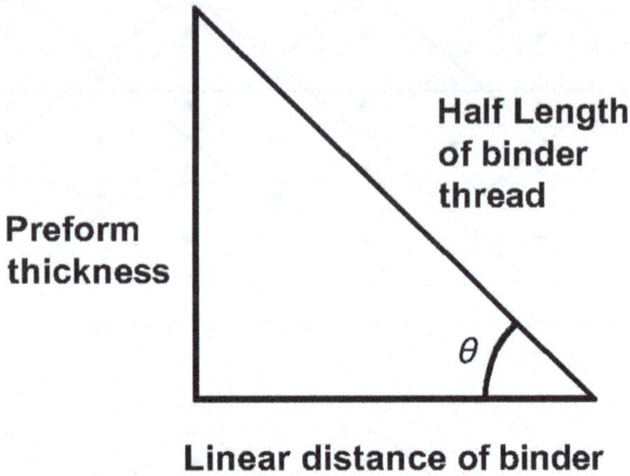

Fig. 14 3D 'T' shape woven using angle interlock weave architecture

sometimes calls for understanding the variations required to obtain the basic parameters such as thickness, fibre weight fraction, X, Y, Z content. From the few samples woven the following can be inferred for 3D angle interlock structures

1. Machine factors: The Warp tension, and reed count implying warp density play their part in affecting the thickness
   Increasing the warp tension would to some extent reduce the preform thickness and of course changing the reed count would in any way change the density. Increasing density would increase the thickness

Table 1 Fibre content calculations of angle interlock woven samples

Details	Thickness (cm)	Stuffer (warp)	Binder (warp)	PPI	Overall fibre volume fraction (%)	'X' content (%)	'Y' content (%)	'Z' content (%)
Sample 1	0.19	2	6	8	46.21	65.59	24.6	9.81
Sample 2	0.2	2 × 2	6	8	51.33	70.13	21.04	8.83
Sample 3	0.2	2	6	12	50.80	55.6	31.26	13.13
Sample 4	0.17	2	6	8	50.97	66.48	29.07	4.45
Sample 5	0.2	2	6	12	51.10	56.36	36.99	6.66
Sample 6	0.22	2	6	16	53.62	48.82	42.72	8.46
Sample 7	0.2	2	4	16	51.33	42.08	49.09	8.83
Sample 8	0.23	2 × 2	4	16	51.49	48.64	42.56	8.81
Sample 9	0.26	4	10	12	55.67	22.36	74.58	3.06

2. Preform factors: Referring to the Figs. 1 to 12, warp density (Ex samples 1,2,3,4), weft density (samples 5,6), Z thread density, yarn count, no. of warp layers (samples 3,4), No. of weft layers, weave architecture, warp arrangement (samples 1,5,8,9) all have a role to play.—Not limiting, it should be understood that Myriad possibilities exist in weaving. The case study of 10 cases here is only representative.

3. Interrelated parameters: Quite a few of the above are interrelated. Ex: The combination of weft density and warp tension will affect the thickness and in turn fibre content.

Calculations for angle interlock based T stiffener: As shown in Fig. 15, the weave architecture comprises of 5 parts. Each part has an individual weave design. Figure 16 shows the close-up of part 2 and part 4 where the technique of bridging is highlighted. Table 2 shows the data obtained from the woven 'T' preform and detailed calculations are performed on the web and flange section separately.

Data obtained from the sample (Base or Flange):

Figure 17 shows a close-up of the flange section shown in Fig. 15

Ends per inch = 8 × 10 = 80.

Picks per inch = 3 × 8 = 24.

Length of the preform = 18 cm.

Width of the preform = 30 cm.

Thickness of the preform = 2 mm = 0.2 cm.

Thread density:

Stuffer warp (X1) = 4 × 8 = 32 per inch (12.6 per cm).

Weft (Y) = 3 × 8 = 24 per inch (9.45 per cm).

The fibre content in X is a combination of stuffer yarn (X1) and Binder yarn (X2).

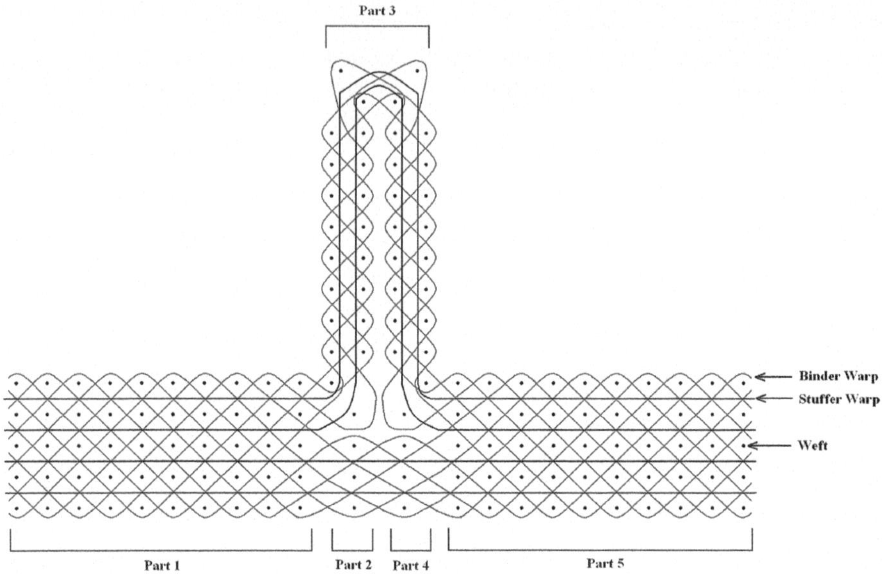

Fig. 15 Weave architecture of angle interlock 'T' shape

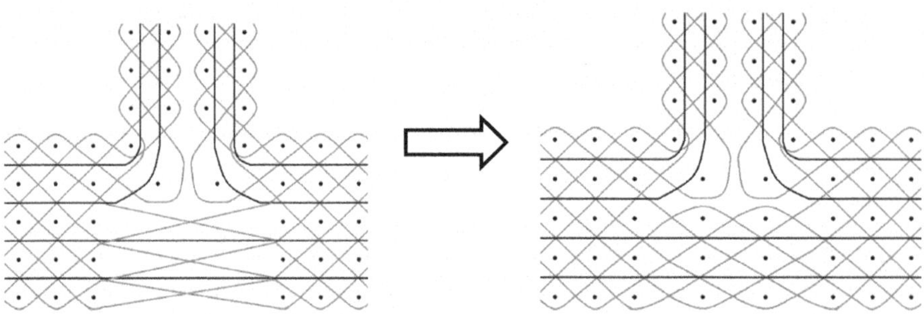

Fig. 16 Flange bridged together by a change in weave architecture

Binder yarn traverses for a length on the face and back of the preform (X2 content) and continues into the thickness portion of the preform (Z content) as shown in Fig. 18.
Binder thread density (X2, Z) (along width) = 6 × 8 = 48 per inch (18.90 per cm).
Carbon yarn count is 400 Tex → 400 g per 1000 m length of yarn for Binder yarns and Stuffer yarns.

Table 2 Woven sample details

Sample details for angle interlock 'T' profile		
Details	Flange	Web
Thickness (cm)	0.26	0.2
Stuffer warp density (per cm)	6.3	3.15
Binder warp density (per cm)	15.75	6.3
Weft density (per cm)	23.62	10.24
Calculated preform weight (g)	141.33	75.2
Fabric volume content (%)	55.67	63.01
Warp content-X (%)	22.36	22.88
Weft content-Y (%)	74.58	74.22
Z content (%)	3.06	2.93
T preform weight (310 mm length)	216.53 g	

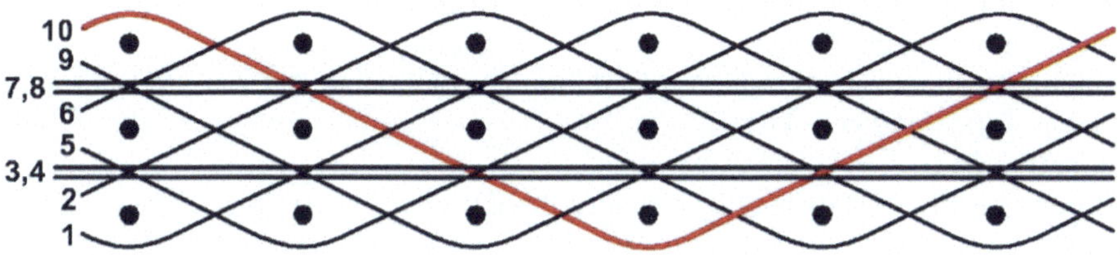

Fig. 17 Flange structure

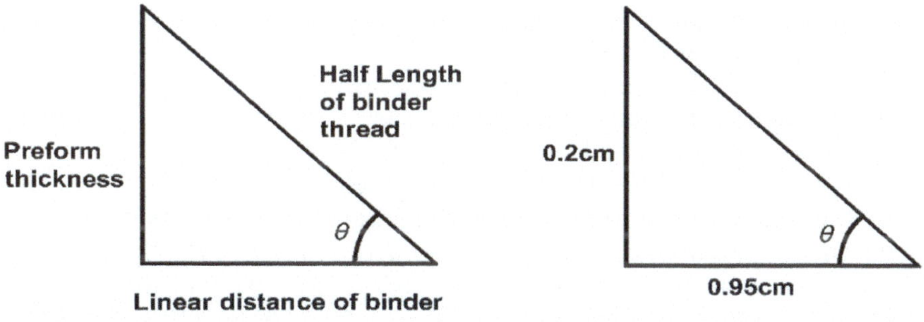

Fig. 18 Vector resolution of the Flange structure

Content of warp (X)

Content of stuffer warp (X1) in the sample

Total number of stuffer warp = Stuffer warp density X sample width = 12.6 X 30 = 378.

Length of stuffer warp = number of stuffer warps X sample length = 378 X 18 = 6804 cm.

Length of stuffer warp in m = 68.04 m.

Carbon yarn count is 400 Tex = > 400 g per 1000 m length of yarn.

Therefore weight of stuffer warp for 12 m = (68.04 X 400) / 1000 = 27.216 g.

Content of Binder warp (X2) in the sample

Since the binder yarn traverses at an angle, it is required to be resolved into X2 component of the binder thread and Z component of the binder thread using vectors.

To calculate the length of binder yarn,

Each binder thread repeats after 6 picks.

Pick density = 8 per inch.

Therefore Linear distance occupied by one binder thread = $(6 / 8) \times 2.54 = 1.91$ cm.

Linear distance occupied by Half binder thread = 1.91 / 2 = 0.95.

Actual length of half binder thread is obtained by Pythagoras Theorem as below.

Half-length of binder thread = $\sqrt{}\ 0.95 = 0.976$ cm.

Length of one binder thread = $0.976 \times 2 = 1.95$ cm.

Each binder thread needs to be resolved into X2 component and Z component.

$\theta = \tan^{-1} (0.2 / 0.95) = 11.89°$.

For half thread,

X2 component = (Half-length of binder thread) $\cos\theta = 0.976 \times \cos(11.89) =$ 0.955.

Z component = (Half-length of binder thread) $\sin\theta = 0.976 \times \sin(11.89) = 0.201$.

X2 component of one binder thread = $0.955 \times 2 = 1.91$ cm.

One repeat of the weave is 8 picks.

X2 component for 6 picks = 1.91 cm.

X2 component for 8 picks = $(1.91 \times 8) / 6 = 2.55$ cm.

X2 component for 1 inch length of preform = 2.542 cm.

X2 component for 1 cm length of preform = 1 cm.

X2 component for 18 cm length of preform = 18 cm.

Number of binder threads for 1 inch width of preform = 48.

Number of binder threads for 1 cm width of preform = 48 / 2.54 = 18.90.

Total binder threads in 30 cm of preform = $30 \times 18.9 = 566.9 \approx 567$.

X2 component of all binder threads (in m) = $(567 \times 18) / 100 = 102.06$.

Weight of X2 component carbon tows of 400 tex = $(400 \times 102.06) / 1000 = 40.824$ g.

Content of warp (X) in the sample = X1 + X2 = 27.21 + 40.82 = 68.03 g.

Content of weft (Y) in the sample

Total number of picks = pick density X sample length = 9.45 X 18 = 170.12.

Length of picks = number of picks X sample width = 170.1 X 30 = 5103 cm = 51.03 m.

Weight of weft for 51.03 m = (51.03 X 400) / 1000 = 20.412 g.

Content of Z threads in the sample

Z component of half binder thread = 0.2.

Z component of one binder thread (for 6 picks) = 0.4.

Z component for 8 picks = (0.4 × 8)/6 = 0.53 cm.

Z component for 1 inch length of preform = 0.53 cm.

Z component for 1 cm length of preform = 0.209 cm.

Z component for 18 cm length of preform = 18 × 0.209 = 3.762 cm.

Z component of all binder thread (in m) = (567 × 3.78) / 100 = 21.43 m.

Weight of Z component of carbon tows of 400 tex = (400/1000) × 21.43 = 8.572 g.

Calculated preform weight = X + Y + Z = 68.03 + 20.412 + 8.532 = 97.01 g.

Calculations for fibre content

Total preform volume = Length X width X height = 18 X 30 X 0.2 = 108 cc.

Fabric content by volume = calculated fabric weight / density of carbon = 97.01 / 1.75 = 55.44 cc.

Total fabric volume fraction = Fabric content by volume / Total preform volume = 55.44/108 = 0.5133 = 51.33%.

Volume of X = Weight of X threads / density of carbon = 68.03/1.75 = 38.87 cc.

Content of warp (X) in preform by volume = Volume of X/ volume of preform = 38.87 / 55.44 = 0.7013 = 70.13%.

Volume of Y = Weight of Y threads / density of carbon = 20.412/1.75 = 11.66 cc.

Content of weft (Y) in preform by v = volume of Y / Volume of preform = 11.66/55.44 = 0.2104 = 21.04%.

Volume of Z = weight of Z threads / density of carbon = 8.532/1.75 = 4.875 cc.

Content of Z threads in preform by volume = volume of Z/ Volume of preform = 4.875/55.44 = 0.0883 = 8.83%.

Data obtained from sample (Web):

Figure 19 is the close-up of the web structure of ‘T’ profile as shown in Fig. 15.

Ends per inch = 6 × 8 = 48.

Picks per inch = 8 × 2 = 16.

Length of the preform = 23 cm.

Width of the preform = 30 cm.

Thickness of the preform = 0.1 cm.

Thread density:

Stuffer warp (X1) = 2 × 4 = 8 per inch (3.15 per cm).

Weft (Y) = 2 × 8 = 16 per inch (6.3 per cm).

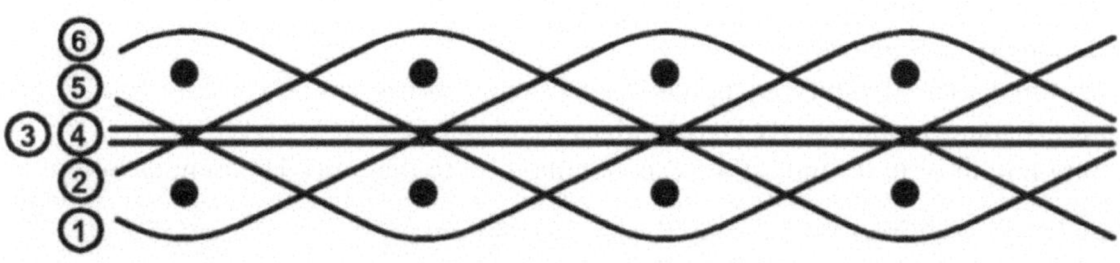

Fig. 19 Web structure

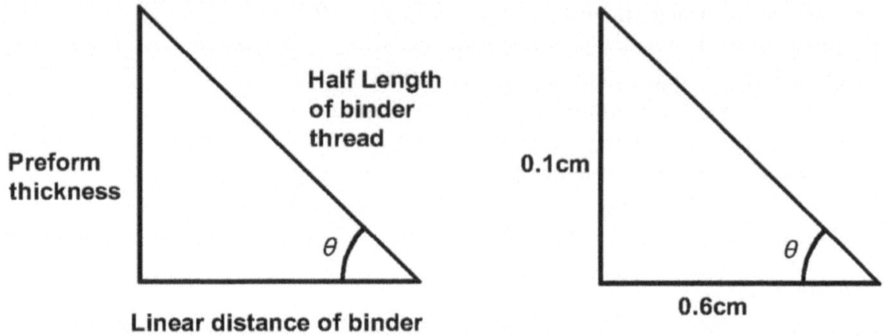

Fig. 20 Vector resolution of web structure

The fibre content in X is a combination of stuffer yarn (X1) and Binder yarn (X2). Binder yarn traverses for a length on the face and back of the preform (X2 content) and continues into the thickness portion of the preform (Z content) as shown in Fig. 20.
Binder thread density (X2, Z) (along width) = $4 \times 8 = 32$ per inch (12.6 per cm).
Carbon yarn count is 400 Tex → 400 g per 1000 m length of yarn for both Binder yarns and Stuffer yarns.

Content of warp (X)

Content of stuffer warp (X1) in the sample.

Total number of stuffer warp = Stuffer warp density X sample width = 3.15 X 30 = 94.5 ≈95 cm.

Length of stuffer warp = number of stuffer warps X sample length = 95 X 23 = 2173.5 cm.
Length of stuffer warp in m = 21.73 m.
Carbon yarn count is 400 Tex = > 400 g per 1000 m length of yarn.
Therefore weight of stuffer warp for 21.73 m = (21.73 X 400) / 1000 = **8.694 g**.

Content of Binder warp (X2) in the sample

Since the binder yarn traverses at an angle, it is required to be resolved into X2 component of the binder thread and Z component of the binder thread using vectors.

To calculate the length of binder yarn,

Each binder thread repeats after 4 picks.
Pick density = 8 per inch.
Therefore Linear distance occupied by one binder thread = $(4/8) \times 2.54 = 1.27$ cm.
Linear distance occupied by Half binder thread = 0.781/2 = 0.635.
Actual length of half binder thread is obtained by Pythagoras Theorem as below.

Half-length of binder thread = $\sqrt{0.413} = 0.642$ cm.
Length of one binder thread = $0.642 \times 2 = 1.29$ cm.
Each binder thread needs to be resolved into X2 component and Z component.

$\theta = \tan^{-1}$ (0.1/ 0.63) = 9.02.
For half thread,
X2 component = (Half-length of binder thread) cos θ = 0.642 × cos (9.02) = 0.634.
Z component = (Half-length of binder thread) sin θ = 0.642 × sin (9.02) = 0.1.
X2 component of one binder thread = 0.634 × 2 = 1.27 cm.
One repeat of the weave is 4 picks.
X2 component for 4 picks = 1.27 cm.
X2 component for 8 picks = (1.27 × 8) / 4 = 2.54 cm.
X2 component for 1 inch length of preform = 2.54 cm.
X2 component for 1 cm length of preform = 1 cm.
X2 component for 23 cm length of preform = 23 cm.
Number of binder threads for 1 inch width of preform = 32.
Number of binder threads for 1 cm width of preform = 32 / 2.54 = 12.6.
Total binder threads in 30 cm of preform = 30 × 12.6 = 377.95 ≈ 378.
X2 component of all binder threads (in m) = (378 × 23) / 100 = 86.92 m.
Weight of X2 component carbon tows of 400 tex = (400 × 86.92) / 1000 = **34.77 g**.
Content of warp (X) in the sample = X1 + X2 = 8.69 + 34.77 = **43.46 g**.

Content of weft (Y) in the sample

Total number of picks = pick density X sample length = 6.3 X 23 = 144.9≈145.

Length of picks = number of picks X sample width = 145 X 30 = 4350 cm = 43.5 m.

Weight of weft for 43.5 m = (43.5 X 400) / 1000 = **17.4 g**.

Content of Z threads in the sample

Z component of half binder thread = 0.1 cm.

Z component of one binder thread (for 4 picks) = 0.2 cm.

Z component for 8 picks = 0.4.

Z component for 1 inch length of preform = 0.4 cm.
Z component for 1 cm length of preform = 0.16 cm.
Z component for 23 cm length of preform = 23 × 0.157 = 3.62 cm.
Z component of all binder thread (in m) = (378 × 3.62) / 100 = 13.64 m.
Weight of Z component of carbon tows of 400 tex = (400/1000) × 13.64 = **5.45 g**.
Calculated preform weight = X + Y + Z = 43.46 + 17.4 + 5.45 = **66.33 g**.

Calculations for fibre content

Total preform volume = Length X width X height = 23 X 30 X 0.1 = 69 cc.

Fabric content by volume = calculated fabric weight / density of carbon = 66.33 /1.75 = 37.9 cc.

Total fabric volume fraction = Fabric content by volume / Total preform volume = 37.9/69 = 0.5428 = 54.93%.

Volume of X = Weight of X threads / density of carbon = 43.46/1.75 = 24.84.
Content of warp (X) in preform by volume = Volume of X/ volume of preform = 24.84/37.9 = 0.6553 = 65.53%.
Volume of Y = Weight of Y threads / density of carbon = 16.632/1.75 = 9.504 cc.

Content of weft (Y) in preform by volume = volume of Y / Volume of preform = 9.504/37.9 = 0.2621 = 26.21%.
Volume of Z = weight of Z threads / density of carbon = 5.458/1.75 = 3.13 cc.
Content of Z threads in preform by volume = volume of Z/ Volume of preform = 3.13/37.9 = 0.08 = 8.26%.

4 Combination of Angle Interlock and Orthogonal Weave Architectures

Each of the orthogonal or angle interlock fabric architectures has unique properties associated with it. For example, angle interlock 3D preform has a good drape and can conform to complex contours easily whereas orthogonal woven fabrics have higher dimensional stability. By combining the two architectures, the scope of designing 3D fabrics with more array of properties increases. The binder yarns of orthogonal weave traverse through the thickness to provide additional reinforcement to the preform structure. Chapter 3 details about the combined structure.

Acknowledgements The following students and their institutes are acknowledged for their contributions vide their project works.

1. Meera R, M Tech, Development and evaluation of 3D angle interlock reinforcements with variation in fibre contents. GSKSJT Institute, Visvesvaraya Technological University, 2014–15.
2. Ashwathi, M Tech, Development and evaluation of 3D angle interlock wovenT shaped preforms for use in composites. GSKSJT Institute, Visvesvaraya Technological University 2013–14.
3. Shwetha Singh.N and Ashwathi. B Tech, Development and evaluation of three dimensional angle interlock woven preform for use in composites. GSKSJT Institute, Visvesvaraya Technological University-2011,12.

References

1. Callus PJ, Mouritz AP and Bannister MK (1999) "Tensile properties and failure mechanisms of 3D woven GRP composites" *Composite Part A,* pp. 1277–1287
2. Chen X, "Mathematical modelling of 3D woven fabrics for CAD/CAM software, Text Res J 1(81)
3. Bilisik K, "New method of weaving multiaxis three dimensional flat woven fabric. Feasibility of Prototype Tube carrier weaving," Fibres and textiles in Eastern Europe
4. Endruweit A, long AC (2010) "Analysis of compressibility and permeability of selected 3D woven reinforcements" J Compos mat 44(24):2833–2862
5. Mahadik Y, Hallett SR (2011) Effect of fabric compaction and yarn waviness on3D woven composite compressive properties. Compos A 42:1592–1600
6. Dai S, Cunningham PR, Marshall S (2015) Influence of Fibre architechure on the tensile, compressive and flexural behavior of 3D woven composites. Composite Part A 59:195–207
7. Lapeyronnie P (2011) phillipe Le Grognec and Christop, "Homogenisation of the elastic behavior of a layer to layer angle interlock composite." Compos Struct 93:2795–2807

8. Archer E, Broderick J, McIlhagger AT (2014) Internal strain measurement and cure monitoring of 3D angle interlock woven carbon fibre composites. Composite Part B 56:424–430
9. Dianyun Z, Anthony MW, Chian FY (2014) "Progressive failure on textile composites" In: 55th AIAA/ASMe/ASCE/AHS/SC Structures, Structural Dynamics, and Materials Conference, Maryland
10. Gras R, Leclere H, Roux S (2013) Identification of the out-of-plane shear modulus of a 3D woven compo-site. Exp Mech 53:719–730
11. Xiagong C, Dan Y "Use of 3D angle-interlock woven fabric for seamless female body armour," *Part I Ballistic Evaluation*
12. Jin L, Sun B, Bohong Gu (2011) Finite Element simulation of three dimensional angle interlock woven fabric undergoing ballistic impact. J Text Inst 102(11):982–993
13. Nawab Y, Legrand X, Koncar V (2012) Study of Changes in 3D woven multilayer interlock fabric preforms while forming. J Text Inst 103(12):1273–1279
14. Sun B, Yao Y, Jin L, Bohong Gu (2013) Finite elment analysis of stress distributions of three dimensional angle interlock woven composite subjected to three point bending cyclic loading. J Text Inst 104(11):1186–1194
15. Yang Ge (2014) Baozhong sun and Bohong Gu, "Large scale finite element analysis of a 3D angle interlock woven composite undergoing low cyclic three point bending fatigue." J Text Inst 105(3):275–293
16. Buchanan S, Grigorash A, Quinn JP (2010) Modelling the geometry of repeat unit cell of the three dimensional weave architectures. J Text Inst 101(7):679–685
17. Walter TR, Subhash G, Shankar BV (2013) A novel method for dynamic short beam shear testing of 3D woven composites. Exp Mech 53:493–503
18. Gregory NM, James D and James K "Effect of fibre architecture on matrix cracking for melt infiltrated SiC/ SIC composite (2010) Int J Appl Ceram Technol 7(3):276–290
19. Cui F, Sun B, Bohong Gu (2010) Fibre inclination model for finite element analysis of three dimensional angle interlock woven composite. J Compos Mater 45(14):1499–1509
20. Jin L, Hong Hu, Sun B, Bohongo Gu (2012) Three point bending fatigue behavior of 3D angle interlock woven composite. J Compos Mater 46(8):883–894
21. Lefebvre M (2013) Boussou Francois and coutellier D, "Influence of high performance yarns degradation inside three dimensional warp interlock fabric." J Ind Text 42(4):475–488
22. Khan K, Umer R (2017) Modeling the viscoelastic compaction response of 3D woven fabrics for liquid composite molding processes. J Reinf Plast Compos 36(18):1299–1315
23. Zhao Y, Jiao Y (2017) Leilei Song and Qian Jiang an, "Influence of fabric architecture and weaving parameter on the thermal conductivities of 3D woven composites." J Compos Mater 51(21):3041–3051
24. Sugun BS, Sandeep DN (2017) "Integral weaving of orthogonal 3D T stiffeners based on pleat weaving concept" J Ind Text pp 1–19
25. Sugun BS, Sandeep DN (2016) Development of single layer 3D 'T' profile with fillet for composite 'T' joints. J Ind Text 46(3):887–900

Chapter 3
Quality Control Aspects for 3D Woven Preforms

B. S. Sugun and Prasad Shimpi

1 Introduction

The weaving industry from time immemorial has an inspection room, wherein, the cloth after doffing from the loom is checked metre by metre for defects and quality. However, this approach is not feasible for quality analysis of 3D preforms, because 3D preforms have a thickness dimension and the interior part cannot be inspected. Also, the fabric construction parameters such as thread densities are calculated in a different way for 3D preforms as it has to account multiple layers as compared to 2D fabric, which has a single layer. This chapter discusses the methods to evaluate quality of 3D fabrics and calculate the fabric construction parameters along with case studies as examples.

Control Parameters.

A typical 3D preform has the following constituents.

X-direction threads.

Y-direction threads.

Z-direction threads (called binder threads).

The X-direction threads are usually regular warp threads, drawn-in through the heald shaft. For Noobed fabric structure, these warp threads are bifurcated into stuffer and binder warps.The stuffer warps are stacked in multiple layers, and they contribute solely to the X direction, whereas, the binder warps act as Z directional threads and contribute a fraction of their length in X direction as well. In the case of angle

B. S. Sugun (✉)
Centre for Societal Missions and Special Technologies (CSMST), CSIR-National Aerospace Laboratory (NAL), Bangalore, India
e-mail: sugun@nal.res.in

P. Shimpi
CSMST, CSIR-NAL (formerly), Bengaluru, India

B. S. Sugun (ed.), *Practical Approach to 3D Weaving*, Materials Horizons: From Nature to Nanomaterials, https://doi.org/10.1007/978-3-030-80424-4_3

Table 1 Control parameters for 3D woven preforms

Established control parameters for 2D fabrics	*Devised control parameters for 3D preforms*	*Acceptable tolerance envisaged for 3D preforms*
Ends per inch	Stuffer warp per inch (for all layers) Binder warps per inch	±5%
Picks per inch	Total picks per inch	±10%
Warp count	Stuffer warp count Binder warp count	±5%
Weft count	Weft count	±5%
Dimensions (L X B)	Dimensions (L X B X H)	±5%
Warp and weft material	Stuffer warp, binder warp and weft material	
Weight per unit area (GSM)	Weight per unit volume (fibre volume content)	
Weave design and architecture	Weave design, architecture and cross-sectional diagrams	
Weave repeat	Picks per cycle of weaving	

interlock structures, there is no bifurcation between stuffer and binder, as each warp thread contributes a part of it's length to X and Z direction respectively. However in some angle interlock cloth constructions, additional stuffer warp are also inserted which contributes solely to X direction. In both the cases, the Y directional threads are always weft. In the case of Dual plane shedding, an added factor would be the crimp of the yarns arising out of the interlacements in X, Y and Z planes.

Starting with basic detailing, Table 1 provides the information to be derived for 3D woven preforms as compared to its equivalent 2D preforms.

For noobed preforms, stuffer warp per inch implies, counting the surface warps using the pick glass and then multiplying it with the number of layers. Total picks per inch imply surface picks per inch as counted by the pick glass multiplied by the number of layers. GSM can still be a valid measure for 3D woven preforms. However, Fibre volume content viz., the volume occupied by the fibres alone in the volumetric space of the cuboid and the X, Y and Z contents by volume or weight would be a pre-requisite. As explained in the first chapter, the 3D weave architecture and cross-sectional diagram is an important requirement prior to weaving. Picks per cycle in 3D weaving give the information of repeat size. For some weave architectures in angle interlock and shaped weaving, this can go to three digit numbers.

2 Notating 3D Preforms

One of the simplest methods of notating 3D preforms can be as below.

NS9B2PPC10PD6 (Typical example).

N—Noobed.

S9—9 layers of Stuffer.
B2—2 layers of binder.
PPC10—Picks per cycle is 10.
PD6—Surface picks per inch is 6.

Based on the above, AITT represents Angle Interlock Through thickness, AILL represents Angle interlock Layer-to-layer and ML represents Multi-Layer. It should be noted that it may not always be appropriate to use the above format of notation and based on the type, the notation may have a different approach.

Ex. Angle Interlock Through Thickness preform (sample 2) discussed below can be notated as AITTWL7PPC98PD8 which comprises of 7 weft layers, Picks per cycle—98 and Pick density—8.

3 Requirements for Arriving at X, Y and Z

For the calculations shown in Chaps. 1 and 2, the following are the requirements.

1. Sample dimensions
2. Thread density
    a. Stuffer warp (X1)
    b. Weft (Y)
    c. Binder warp (along width) (X2)
    d. Z fibre (along length)—where relevant such as Noobing
3. Material details
    a. Count of stuffer warp
    b. Count of binder warp
    c. Count of weft
    d. Material density

The practical verification of the calculations is by weight measurement of the preform which should be within 10% of the calculated weight. Based on the end use requirement, this can be reduced to 5%.

4 Checking of Weaving Uniformity

For a 2D fabric, thread density at multiple places gives an indication of construction uniformity. However, Surface thread density measurement for 3D fabrics may not suffice to account for weaving uniformity as information within the layers will not be decipherable so easily. One thought out practical approach for this, has been, the Air permeability testing as per ASTM D737 randomly at multiple places (atleast 10) of the preform. The result obtained can be analysed for weaving uniformity.

Table 2 Air Permeability test result of sample NS9B2PPC10PD8

Test details Air permeability test as per ASTM D737 Pressure 125 Pa Test area: 38 cm/Sq No. of test points: 10 Location of points: randomized on the preform	
Sl No.	Air permeability value
1	65.7
2	54.9
3	65.8
4	49.8
5	50.8
6	54.9
7	64.7
8	54.3
9	51.8
10	64.3
Avg	**57.7**
Sd	**6.6**
CoV	**11.4**

In one typical study carried out as part of this evaluation, air permeability test result of a Noobed sample (NS9B2PPC10PD8) has been detailed in Table 2.

5 Concise Data Sheets and Weave Designs as Woven on a 24 Shaft Dobby Loom for Different Types of 3D Woven Preforms (Noobed, Angle Interlock, Combination)

For the reader to understand and appreciate the intricacies involved in designing, weaving and arriving at the requisite calculation, the following samples along with their data sheets have been provided for reference. This is apart from the examples discussed in Chaps. 1 and 2. The five samples that are discussed here are as follows. Noobed-Sample 1 (Figs. 1, 2, 3 and Tables 3, 4, 5, 6), Through thickness angle interlock-Sample 2 (Figs. 4, 5, 6 and Tables 7, 8, 9, 10), Layer-to-Layer angle interlock-Sample 3 (Figs. 7, 8, 9 and tables 11, 12, 13). Combined Noobed & angle interlock (Figs. 10, 11, 12 and tables 14, 15, 16, 17, 18, 19) and multilayer preform (Figs. 13, 14, 15 and tables 20, 21, 22). Other then the dual plane shedding based 3D woven preform, the above 5 samples represent the spectrum of 3D structures that can be basically developed and on keen understanding of the calculations the distinction in them can be understood.

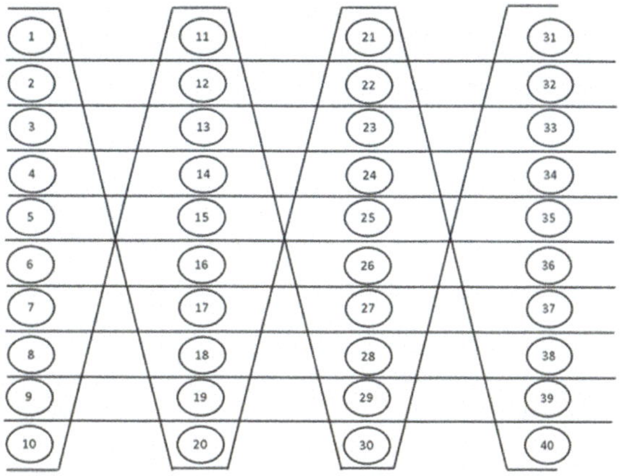

Fig. 1 Weave cross-section (weft way) of simple Noobed preform (sample 1)

No.	1	2	3	4	5	6	7	8	9	10	11	12	13	14	15	16	17	18	19	20	21	22	23	24	PPI	NOP	Beat
22	X	X	X	X	X	X	X	X	X	X	X														8	1	1
21		X																							8	0	1
20		X	X																						8	0	1
19		X	X	X																					8	0	1
18		X	X	X	X																				8	0	1
17		X	X	X	X	X																			8	0	1
16		X	X	X	X	X	X																		8	0	1
15		X	X	X	X	X	X	X																	8	0	1
14		X	X	X	X	X	X	X	X																8	0	1
13		X	X	X	X	X	X	X	X	X															8	0	1
12		X	X	X	X	X	X	X	X	X	X														8	0	1
11	X	X	X	X	X	X	X	X	X	X	X														8	1	1
10	X		X	X	X	X	X	X	X	X	X														8	0	1
9	X		X	X	X	X	X	X	X	X															8	0	1
8	X		X	X	X	X	X	X	X																8	0	1
7	X		X	X	X	X	X	X																	8	0	1
6	X		X	X	X	X	X																		8	0	1
5	X		X	X	X	X																			8	0	1
4	X		X	X	X																				8	0	1
3	X		X	X																					8	0	1
2	X		X																						8	0	1
1	X																								8	0	1

Fig. 2 Weave design as woven on the loom (sample 1)

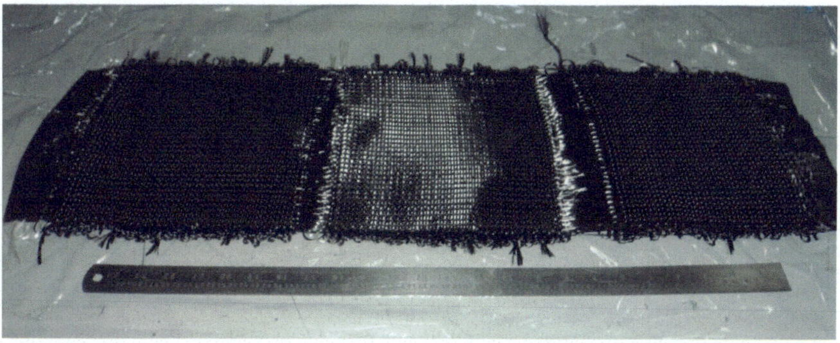

Fig. 3 Photograph of simple noobed preforms (3 nos woven in series (sample 1))

Table 3 Summarized data sheet of simple noobed preform (sample 1)

1. Preform code 2. Raw Material details	NS9B2PPC10PD8 (noobed preform)		
a. Material b. Source c. Trade Name d. Type (Twisted or Untwisted) e. Count f. Modulus	Carbon tows Tairyfil, Korea TC 33 Untwisted 400 Tex Standard—230 GPa		
3 Preparatory process details Warping (beam or creel) Total No. of stuffer warp Total No. of binder warp	Creel 324 72		
4. Preform details	**As planned**	**Actual obtained**	
Overall dimensions (L × B × H) in Cm	23 × 23 × 0.3	23 × 22 × 0.3	
Total ends per inch for stuffer = No. of warp Layers × Surface ends/inch	9 × 4 = 36	9 × 4 = 36	
Binder threads per inch = No. of ends per dent × No. of binder dents/inch	8	8	
Total picks per inch = Picks per cycle × surface picks per inch	10 × 8 = 80	10 × 8 = 80	
Preform weight	104.83	108	Weft loops
Total fibre volume content	39.46		
X content (%)	33.45		
Y content (%)	60.81		
Z content(%)	5.75		
Weave cross-section and weave design as woven on the loom	Figs. 1, 2		
Summarized calculations of X, Y and Z content and fibre volume content	Tables 4, 5 and 6		
Sample photographs	Fig. 3		

Table 4 Data obtained from preform (sample 1)

Preform length in cm	23	
Preform width in cm	22	–
Preform thickness in cm	0.3	–
Thread density	per cm	Per inch
Stuffer warp (X1)	14.17	36
Weft (Y)	31.50	80
Binder warp (along width)—X2	3.15	8
Z fibre (along length)	3.15	8
Carbon fibre details		
Stuffer warp (X1) in tex	400	6 K
Weft (Y) in tex	400	6 K
Binder warp in tex	400	6 K
Carbon fibre density in g/cc	1.75	–

Table 5 Calculations (summarized) for X, Y and Z contents (sample 1)

Content of warp (X)	
Content of stuffer warp (X1)	–
Total number of stuffer warp	312
Total length of stuffer warp in mtr	71.72
Weight of stuffer warp in gms	28.69
Content of Binder warp (X2)	–
Total number of binder warp	69
Total length of binder warp in mtr	15.94
Weight of binder warp in gms	6.37
Content of warp (X) in gms	**35.06**
Content of weft (Y)	
Total number of wefts	724
Total length of weft in 1 mtr	159.37
Weight of weft in gms	63.75
Content of weft (Y) in gms	**63.75**
Content of Z threads	
Total number of binder warp	69
Number of Z threads for one binder thread	72.44
Total length of Z threads	15.06
Weight of Z threads in gms	6.02

Table 6 Fibre volume fraction calculations and preform weight (sample 1)

Fibre volume fraction calculations		
Total preform volume in cc	151.80	–
Total preform fibre content by volume	58.18	–
Total preform fibre volume fraction	**0.39**	**39.46%**
Volume of X fibres	20.04	–
Content of warp—X	**0.33**	**33.45%**
Volume of Y fibres	36.43	–
Content of weft—Y	0.61	**60.81%**
Volume of Z fibres	3.44	–
Content of Z thread	**0.06**	5.75%
Total weight of the preform (individual)	**104.83 gms**	

Fig. 4 Weave cross-section (weft way) of through thickness Angle Interlock Preform (sample 2)

a

No.	1	2	3	4	5	6	7	8	9	10	11	12	13	14	15	16	17	18	19	20	21	22	23	24	PPI	NOP	Beat
43						X	X	X	X	X	X	X	X	X											8	0	1
42	X				X	X	X	X	X	X	X	X	X	X											8	0	1
41	X	X		X	X	X	X	X	X	X	X	X	X	X											8	0	1
40	X	X	X	X	X	X	X	X	X	X	X	X	X	X											8	1	1
39	X	X	X		X	X	X	X	X	X	X	X	X	X											8	0	1
38	X	X				X	X	X	X	X	X	X	X	X											8	0	1
37	X					X	X	X	X	X	X	X		X											8	0	1
36						X	X	X	X	X	X	X													8	0	1
35							X	X	X	X	X														8	0	1
34								X	X	X															8	0	1
33									X							X									8	0	1
32	X	X	X	X	X	X	X	X	X	X	X	X	X	X											8	1	1
31								X								X									8	0	1
30							X	X	X																8	0	1
29						X	X	X	X	X															8	0	1
28	X					X	X	X	X	X	X														8	0	1
27	X	X				X	X	X	X	X	X	X													8	0	1
26	X	X	X			X	X	X	X	X	X	X		X											8	0	1
25	X	X	X	X		X	X	X	X	X	X	X	X	X											8	0	1
24	X	X	X	X	X	X	X	X	X	X	X	X	X	X											8	1	1
23	X	X	X	X	X	X	X	X	X	X	X	X		X											8	0	1
22	X	X	X	X		X	X	X	X	X	X	X													8	0	1

Fig. 5 **a** Weave design for picks 22–43 (sample 2). **b** Weave design for picks 1–21 (sample 2). **c** Weave design for picks 44–65 (sample 2). **d** Weave design for picks 66–87 (sample 2). **e** Weave design for picks 88–109 (sample 2). **f** Weave design for picks 110–112 (sample 2). (**a** through **f**) Weave design as woven on the loom (sample 2)

b

No.	1	2	3	4	5	6	7	8	9	10	11	12	13	14	15	16	17	18	19	20	21	22	23	24	PPI	NOP	Beatu_
21	X	X	X			X	X	X	X	X	X														8	0	1
20	X	X				X	X	X	X	X															8	0	1
19	X					X	X	X	X																8	0	1
18						X	X	X																	8	0	1
17							X									X									8	0	1
16	X	X	X	X	X	X	X	X	X	X	X	X	X	X											8	1	1
15						X										X									8	0	1
14	X					X	X																		8	0	1
13	X	X				X	X	X																	8	0	1
12	X	X	X			X	X	X	X																8	0	1
11	X	X	X	X		X	X	X	X	X															8	0	1
10	X	X	X	X	X	X	X	X	X	X	X														8	0	1
9	X	X	X	X	X	X	X	X	X	X	X	X	X												8	0	1
8	X	X	X	X	X	X	X	X	X	X	X	X	X	X											8	1	1
7	X	X	X	X	X	X	X	X	X	X	X		X	X											8	0	1
6	X	X	X	X	X	X	X	X	X	X			X												8	0	1
5	X	X	X	X	X	X	X	X	X																8	0	1
4	X	X	X	X		X	X	X																	8	0	1
3	X	X	X			X	X																		8	0	1
2	X	X				X																			8	0	1
1	X															X									8	0	1

c

No.	1	2	3	4	5	6	7	8	9	10	11	12	13	14	15	16	17	18	19	20	21	22	23	24	PPI	NOP	Beatu_
65														X		X									8	0	1
64	X	X	X	X	X	X	X	X	X	X	X	X	X	X											8	1	1
63												X				X									8	0	1
62											X	X		X											8	0	1
61										X	X	X	X	X											8	0	1
60					X				X	X	X	X	X	X											8	0	1
59				X	X			X	X	X	X	X	X	X											8	0	1
58			X	X	X		X	X	X	X	X	X	X	X											8	0	1
57		X	X	X	X	X	X	X	X	X	X	X	X	X											8	0	1
56	X	X	X	X	X	X	X	X	X	X	X	X	X	X											8	1	1
55	X		X	X	X	X	X	X	X	X	X	X	X	X											8	0	1
54				X	X	X	X	X	X	X	X	X	X	X											8	0	1
53					X		X	X	X	X	X	X	X	X											8	0	1
52								X	X	X	X	X	X	X											8	0	1
51									X	X	X	X		X											8	0	1
50										X	X	X													8	0	1
49											X					X									8	0	1
48	X	X	X	X	X	X	X	X	X	X	X	X	X	X											8	1	1
47										X						X									8	0	1
46									X	X	X														8	0	1
45								X	X	X	X	X													8	0	1
44							X	X	X	X	X	X		X											8	0	1

Fig. 5 (continued)

d

No.	1	2	3	4	5	6	7	8	9	10	11	12	13	14	15	16	17	18	19	20	21	22	23	24	PPI	NOP	Beat
87	X	X	X	X	X	X	X		X	X	X	X	X	X											8	0	1
86	X	X	X	X	X	X				X	X	X	X	X											8	0	1
85	X	X	X	X	X						X	X	X	X											8	0	1
84		X	X	X	X							X	X	X											8	0	1
83			X	X	X								X	X											8	0	1
82				X	X								X												8	0	1
81					X											X									8	0	1
80	X	X	X	X	X	X	X	X	X	X	X	X	X	X											8	1	1
79													X			X									8	0	1
78					X								X	X											8	0	1
77				X	X							X	X	X											8	0	1
76			X	X	X						X	X	X	X											8	0	1
75		X	X	X	X					X	X	X	X	X											8	0	1
74	X	X	X	X	X				X	X	X	X	X	X											8	0	1
73	X	X	X	X	X	X		X	X	X	X	X	X	X											8	0	1
72	X	X	X	X	X	X	X	X	X	X	X	X	X	X											8	1	1
71	X	X	X	X	X		X	X	X	X	X	X	X	X											8	0	1
70		X	X	X	X			X	X	X	X	X	X	X											8	0	1
69			X	X	X				X	X	X	X	X	X											8	0	1
68				X	X					X	X	X	X	X											8	0	1
67					X						X	X	X	X											8	0	1
66												X	X	X											8	0	1

e

No.	1	2	3	4	5	6	7	8	9	10	11	12	13	14	15	16	17	18	19	20	21	22	23	24	PPI	NOP	Beat
109	X	X	X	X		X																			8	0	1
108	X	X	X	X	X	X	X																		8	0	1
107	X	X	X	X	X	X	X	X					X												8	0	1
106	X	X	X	X	X	X	X	X	X				X	X											8	0	1
105	X	X	X	X	X	X	X	X	X	X		X	X	X											8	0	1
104	X	X	X	X	X	X	X	X	X	X	X	X	X	X											8	1	1
103	X	X	X	X	X	X	X	X	X		X	X	X	X											8	0	1
102	X	X	X	X	X	X	X	X				X	X	X											8	0	1
101	X	X	X	X	X	X	X						X	X											8	0	1
100	X	X	X	X	X	X							X												8	0	1
99	X	X	X	X	X																				8	0	1
98		X	X	X																					8	0	1
97			X													X									8	0	1
96	X	X	X	X	X	X	X	X	X	X	X	X	X	X											8	1	1
95				X												X									8	0	1
94			X	X	X																				8	0	1
93		X	X	X	X								X												8	0	1
92	X	X	X	X	X								X	X											8	0	1
91	X	X	X	X	X	X						X	X	X											8	0	1
90	X	X	X	X	X	X	X				X	X	X	X											8	0	1
89	X	X	X	X	X	X	X	X		X	X	X	X	X											8	0	1
88	X	X	X	X	X	X	X	X	X	X	X	X	X	X											8	1	1

f

No.	1	2	3	4	5	6	7	8	9	10	11	12	13	14	15	16	17	18	19	20	21	22	23	24	PPI	NOP	Bea
112	X	X	X	X	X	X	X	X	X	X	X	X	X	X											8	1	1
111		X														X									8	0	1
110	X	X	X																						8	0	1

Fig. 5 (continued)

Fig. 6 Photograph of angle interlock through thickness preform (sample 2)

Table 7 Summarized data sheet of through thickness angle interlock preform (sample 2)

1. Preform code	AITTWL7PPC98PD8 Angle interlock through thickness with 7 weft layers, picks per cycle—98 and pick density—8	
2. Raw material details a. Material b. Source c. Trade name d. Type (Twisted or Untwisted) e. Count f. Modulus	Carbon tows Tairyfil, Korea TC 33 Untwisted 400 tex Standard—230 GPa	
3. Preparatory process details Warping (beam or creel) Total no. of binder warp Total no. of stuffer warp	Creel 504 0	
4. Preform details	**As calculated**	**Actual Obtained**
Overall dimensions (L × B × H) in Cm	23 × 23 × 0.3	24 × 22 × 0.3
Total ends per inch for binder = No. of warp layers × surface ends/inch	7 × 8 = 56	7 × 8 = 56
Stuffer threads per inch = No. of ends per dent × No. of binder dents/inch	NA	NA
Total picks per inch = Picks per cycle × surface picks per inch	7 × 8 = 56	7 × 8 = 56
Preform weight	100.66	100
Total fibre volume content	36.07	
X content (%)	48.39	
Y content (%)	46.57	
Z content (%)	5.71	
Weave cross-section and weave design as woven on the loom	Fig. 4, 5	
Summarized calculations of X, Y and Z content and fibre volume content	Tables 8, 9 and 10	
Sample photograph	Fig. 6	

Table 8 Data obtained from preform (sample 2)

Preform length in cm	24	–
Preform width in cm	22	–
Preform thickness in cm	0.3	–
Thread density	Per Cm	Per inch
Weft (Y)	3.15	8
Binder warp (along width)—X	3.15	8
Carbon fibre details	–	–
Binder warp (X) in tex	400	6 K
Weft (Y) in tex	400	6 K
Carbon fibre density in g/cc	1.75	–

Table 9 Calculations (summarized) for X, Y and Z contents (sample 2)

Content of warp (X)	
Content of binder warp (X)	–
No. of pick columns on which binder thread repeats	14
No. of pick columns per inch	8
Linear distance occupied by 1/2 binder	2.54 cm
Half-length of binder thread	2.558 cm
Angle in degree	6.73
X component of 1/2 binder thread	2.54 cm
X component for 24 cm length preform	24 cm
X component for total preform length	120.96 m
Total number of binder warp	504
Total length of binder warp in mtr	121.7
Weight of binder warp in gm	48.68 gm
Content of warp (X) in gm	**48.384 gm**
Content of weft	
Total number of wefts	529
Total length of weft in mtr	116.42
Weight of weft in gm	46.57
Content of weft (Y) in gm	**46.57**
Content of Z threads	
Total number of binder warp	504
Number of Z threads for half binder thread	0.3
Total length of Z threads	14.28 m
Weight of Z threads in gm	5.712
Content of Z threads in gm	**5.712**

Table 10 Fibre volume fraction (sample 2)

Total preform volume in cc	158.4	
Total preform fibre content by volume	58.82	–
Total preform fibre volume fraction	**0.36**	**36.07%**
Volume of X fibres	27.65	–
Content of warp—X	**0.48**	**48.39%**
Volume of Y fibres	26.61	–
Content of weft—Y	**0.47**	**46.57%**
Volume of Z fibres	3.26	–
Content of Z thread	**0.06**	**5.71%**
Total weight of the preform	**100.66 gms**	

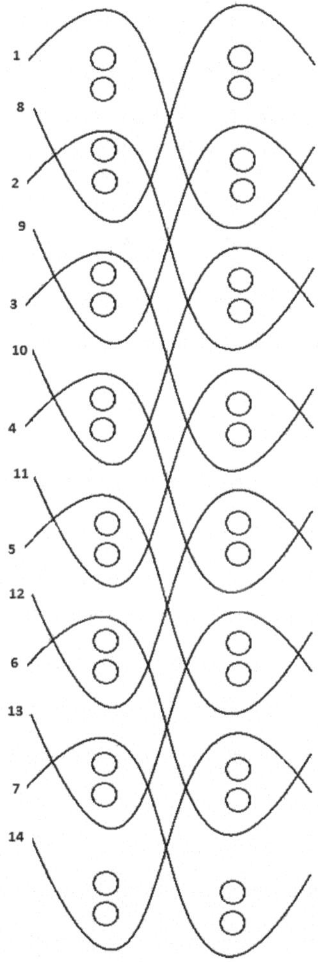

Fig. 7 Weave cross-section (weft way) of layer-to-layer angle interlock preform (sample 3)

No.	1	2	3	4	5	6	7	8	9	10	11	12	13	14	15	16	17	18	19	20	21	22	23	24	PPI	NOP	Beat
32	X	X	X	X	X	X	X	X	X	X	X	X	X	X											8	1	1
31						X								X	X	X									8	0	1
30						X								X	X	X									8	0	1
29						X	X																		8	0	1
28						X	X																		8	0	1
27	X					X	X	X																	8	0	1
26	X					X	X	X																	8	0	1
25	X	X				X	X	X	X																8	0	1
24	X	X				X	X	X	X																8	0	1
23	X	X	X			X	X	X	X	X															8	0	1
22	X	X	X			X	X	X	X	X															8	0	1
21	X	X	X	X		X	X	X	X	X	X														8	0	1
20	X	X	X	X		X	X	X	X	X	X														8	0	1
19	X	X	X	X	X	X	X	X	X	X	X			X											8	0	1
18	X	X	X	X	X	X	X	X	X	X	X			X											8	0	1
17	X	X	X	X	X	X	X	X	X	X	X	X		X											8	0	1
16	X	X	X	X	X	X	X	X	X	X	X	X		X											8	0	1
15	X	X	X	X	X	X	X	X	X	X	X	X	X	X											8	1	1
14	X	X	X	X	X	X	X	X	X	X		X	X												8	0	1
13	X	X	X	X	X	X	X	X	X	X		X	X												8	0	1
12	X	X	X	X	X	X	X	X	X			X													8	0	1
11	X	X	X	X	X	X	X	X	X			X													8	0	1
10	X	X	X	X	X	X	X	X																	8	0	1
9	X	X	X	X	X	X	X	X																	8	0	1
8	X	X	X	X		X	X																		8	0	1
7	X	X	X	X		X	X																		8	0	1
6	X	X	X			X																			8	0	1
5	X	X	X			X																			8	0	1
4	X	X																							8	0	1
3	X	X																							8	0	1
2	X																								8	0	1
1	X																								8	0	1

Fig. 8 Weave design as woven on the loom (sample 3)

Fig. 9 Photograph of woven sample (sample 3)

Table 11 Summarized data sheet of layer-to-layer angle interlock preform (sample 3)

1. Preform code 2. Raw material details	AILLWL8W16PPC32PD6 (8 weft layers,16 weft, Picks per cycle—32, Pick density—6)		
a. Material b. Source c. Trade name d. Type (twisted or untwisted) e. Count f. Modulus	Carbon tows Tairyfil, Korea TC 33 Untwisted 400 Tex Standard—230 GPa		
3. Preparatory process details Warping (beam or creel) Total No. of binder warp Total No. of stuffer warp	Creel 504 0		
4. Preform details	**As calculated**	**Actual Obtained**	
Overall dimensions (L × B × H) in Cm	23 × 23 × 0.3	22.5 × 22 × 0.3	
Total ends per inch for binder = No. of warp layers × surface ends/inch	7 × 8 = 56	7 × 8 = 56	
Stuffer threads per inch = No. of ends per dent × No. of binder dents/inch	NA	NA	

(continued)

Table 11 (continued)

1. Preform code 2. Raw material details	AILLWL8W16PPC32PD6 (8 weft layers,16 weft, Picks per cycle—32, Pick density—6)		
Total picks per inch = Picks per cycle × surface picks per inch	16 × 6 = 96	16 × 6 = 96	
Preform weight	125.65	130	Side weft loops
Total fibre volume content	0.46		
X content (%)	36.74%		
Y content (%)	63.26%		
Z content (%)	NIL		
Weave cross-section and Weave design as woven on the loom	Figs. 7, 8		
Summarized calculations of X, Y and Z content and fibre volume content	Tables 12, 13		
Sample photograph	Fig. 9		

Table 12 Data obtained from preform (sample 3)

Preform length in cm	22.5	
Preform width in cm	23	–
Preform thickness in cm	0.3	–
Thread density	Per cm	Per inch
Weft (Y)	37.80	96
Binder warp (along width)—X	22.05	56
Carbon fibre details		
Binder warp (X) in tex	400	6 K
Weft (Y) in tex	400	6 K
Carbon fibre density in g/cc	1.75	–

Calculations (summarized) X, Y and Z contents and Fibre volume content.
Content of warp (X).
7 Ends/dent, so 56 Ends/inch (16's Reed) Width of Fabric

Total fabric width = 9 inches => 504 ends,

With crimp of 2% and length of preform 225 mm, length of 1 thread= 229 mm. Total length of warp = 504 × 229 = 115416 = 115.416 m ≈ **46.16** g.

Content of weft (Y).

Picks per inch = 16 (through thickness weft layers) × 6 (surface picks/inch) = 96.

Total picks in 9" length of fabric = 96 × 9 = 864.

Total length of weft in 230 mm width of fabric = 864 × 230 = 198.72 m ≈ **79.49 g.**

Table 13 Fibre volume fraction (sample 3)

Total preform volume in cc	155.25	
Total preform fibre content by volume	71.80	–
Total preform fibre volume fraction	**0.46**	**46.25%**
Volume of X fibres	26.38	–
Content of warp—X	**0.37**	**36.74%**
Volume of Y fibres	45.42	–
Content of weft—Y	**0.6323**	**63.26%**
Total weight of the preform	**125.65 gms**	

Fig. 10 Weave cross-section (weft way)of combined noobed and angle interlock preform (sample 4)

a

No.	1	2	3	4	5	6	7	8	9	10	11	12	13	14	15	16	17	18	19	20	21	22	23	24	PPI	NOP	Beat
21	X	X	X	X	X	X	X	X	X	X	X	X	X	X											12	1	1
20	X	X	X	X	X	X	X	X	X	X	X			X											12	0	1
19	X	X	X	X		X	X	X	X	X	X														12	0	1
18	X	X	X			X	X	X	X	X															12	0	1
17	X	X				X	X	X	X																12	0	1
16	X					X	X	X																	12	0	1
15	X						X																		12	0	1
14	X	X	X	X	X	X	X	X	X	X	X	X	X	X											12	1	1
13						X						X													12	0	1
12		X				X	X					X													12	0	1
11		X	X			X	X	X				X													12	0	1
10		X	X	X		X	X	X	X			X													12	0	1
9		X	X	X	X	X	X	X	X	X		X													12	0	1
8		X	X	X	X	X	X	X	X	X	X	X	X												12	0	1
7	X	X	X	X	X	X	X	X	X	X	X	X	X	X											12	1	1
6		X	X	X	X	X	X	X	X	X		X	X	X											12	0	1
5		X	X	X	X	X	X	X	X			X	X												12	0	1
4		X	X	X	X	X	X	X				X													12	0	1
3		X	X	X		X	X					X													12	0	1
2		X	X			X						X													12	0	1
1		X										X													12	0	1

b

No.	1	2	3	4	5	6	7	8	9	10	11	12	13	14	15	16	17	18	19	20	21	22	23	24	PPI	NOP	Beat
43	X										X														12	0	1
42	X	X	X	X	X	X	X	X	X	X	X	X	X	X											12	1	1
41										X		X													12	0	1
40									X	X	X	X													12	0	1
39								X	X	X	X	X		X											12	0	1
38							X	X	X	X	X	X	X	X											12	0	1
37					X	X	X	X	X	X	X	X	X	X											12	0	1
36		X		X	X	X	X	X	X	X	X	X	X	X											12	0	1
35	X	X	X	X	X	X	X	X	X	X	X	X	X	X											12	1	1
34		X	X		X	X	X	X	X	X	X	X	X	X											12	0	1
33		X				X	X	X	X	X	X	X	X	X											12	0	1
32						X	X	X	X	X	X	X		X											12	0	1
31							X	X	X	X	X	X													12	0	1
30								X	X	X		X													12	0	1
29									X			X													12	0	1
28	X	X	X	X	X	X	X	X	X	X	X	X	X	X											12	1	1
27	X							X																	12	0	1
26	X						X	X	X																12	0	1
25	X					X	X	X	X	X															12	0	1
24	X	X				X	X	X	X	X	X														12	0	1
23	X	X	X			X	X	X	X	X	X			X											12	0	1
22	X	X	X	X		X	X	X	X	X	X		X	X											12	0	1

Fig. 11 **a** Weave design for Picks 1–21 (sample 4). **b** Weave design for Picks 22–43 (sample 4). **c** Weave design for Picks 44–65 (sample 4). **d.** Weave design for picks 63–84 (sample 4) (**a**–**d**) Weave design as woven on the loom (sample 4)

c

No.	1	2	3	4	5	6	7	8	9	10	11	12	13	14	15	16	17	18	19	20	21	22	23	24	PPI	NOP	Beat
65		X	X	X	X	X				X	X	X	X	X											12	0	1
64		X	X	X	X	X	X		X	X	X	X	X	X											12	0	1
63	X	X	X	X	X	X	X	X	X	X	X	X	X	X											12	1	1
62		X	X	X	X	X		X	X	X	X	X	X	X											12	0	1
61		X	X	X	X				X	X	X	X	X	X											12	0	1
60			X	X	X					X	X	X	X	X											12	0	1
59				X	X						X	X	X	X											12	0	1
58					X							X	X	X											12	0	1
57												X	X												12	0	1
56	X	X	X	X	X	X	X	X	X	X	X	X	X	X											12	1	1
55	X													X											12	0	1
54	X										X		X	X											12	0	1
53	X				X					X	X		X	X											12	0	1
52	X			X	X				X	X	X		X	X											12	0	1
51	X		X	X	X			X	X	X	X		X	X											12	0	1
50	X	X	X	X	X		X	X	X	X	X		X	X											12	0	1
49	X	X	X	X	X	X	X	X	X	X	X	X	X	X											12	1	1
48	X		X	X	X	X	X	X	X	X	X		X	X											12	0	1
47	X			X	X		X	X	X	X	X		X	X											12	0	1
46	X				X			X	X	X	X		X	X											12	0	1
45	X								X	X	X		X	X											12	0	1
44	X									X	X			X											12	0	1

d

No.	1	2	3	4	5	6	7	8	9	10	11	12	13	14	15	16	17	18	19	20	21	22	23	24	PPI	NOP	Beat
84	X	X	X	X	X	X	X	X	X	X	X	X	X	X											12	1	1
83	X		X																						12	0	1
82	X	X	X	X																					12	0	1
81	X	X	X	X	X	X																			12	0	1
80	X	X	X	X	X	X	X						X												12	0	1
79	X	X	X	X	X	X	X	X					X	X											12	0	1
78	X	X	X	X	X	X	X	X	X		X		X	X											12	0	1
77	X	X	X	X	X	X	X	X	X	X	X	X	X	X											12	1	1
76	X	X	X	X	X	X	X	X		X	X		X	X											12	0	1
75	X	X	X	X	X	X	X				X		X	X											12	0	1
74	X	X	X	X	X	X							X	X											12	0	1
73	X	X	X	X	X								X												12	0	1
72	X		X	X	X																				12	0	1
71	X			X																					12	0	1
70	X	X	X	X	X	X	X	X	X	X	X	X	X	X											12	1	1
69					X							X													12	0	1
68				X	X							X	X												12	0	1
67			X	X	X							X	X	X											12	0	1
66		X	X	X	X						X	X	X	X											12	0	1
65		X	X	X	X	X				X	X	X	X	X											12	0	1
64		X	X	X	X	X	X		X	X	X	X	X	X											12	0	1
63	X	X	X	X	X	X	X	X	X	X	X	X	X	X											12	1	1

Fig. 11 (continued)

Fig. 12 Photograph of combined noobed and angle interlock preform (sample 4)

Table 14 Summarized data sheet of combined noobed/angle interlock preform (sample 4)

1. Preform code	CNAIWL6NB2AIB12PPC72PD10 Combined noobed angle interlock preform—6 weft layers, 2 binders (noobed),12 binders (angle interlock) picks Per cycle—72, pick density—10		
a. Material b. Source c. Trade name d. Type e. count f. Modulus	Carbon tows Tairyfil, Korea TC 33 Untwisted 400 tex Standard—230 GPa		
3. Preparatory process details Warping (beam or creel) Total No. of stuffer warp Total No. of binder warp	Creel 0 504		
4. Preform details	**As Calculated**	**Actual Obtained**	
Overall dimensions (L × B × H) in Cm	20 × 23 × 0.3	20 × 23 × 0.25	
Total ends per inch for stuffer = No. of warp Layers × Surface ends/inch	0	0	
Binder threads per inch = No. of ends per dent × No. of binder dents/inch	7 × 8 = 56	7 × 8 = 56	
Total picks per inch = Picks p er cycle × surface picks per inch	6 × 9 = 54	6 × 9 = 54	
Preform weight	86.94	85	Irregularities in weft density and decimal error
Total fibre volume content	43.20		
X content (%)	Xn + Xa = 5.80 + 34.77 = 40.57/46.66%		
Y content (%)	39.12 / 45%		
Z content (%)	Zn + Za = 2.85 + 4.40 = 7.25/8.34%		
Weave cross-section, Weave designs	Fig. 10, 11		
Calculations	Tables 15, 16, 17, 18, 19		
Photograph	Fig. 12		

Table 15 Data obtained from preform—noobed part (sample 4)

Preform length in cm	20	
Preform width in cm	23	–
Preform thickness in cm	0.25	–
Tread density	Per Cm	Per inch
Stuffer warp (X1)	0	0
Weft (Y)	21.26	54
Binder warp (along width)—X2	3.15	8
Z fibre (along length)	1.97	5
Carbon fibre details		
Stuffer warp (X1) in tex	400	6 K
Weft (Y) in tex	400	6 K
Binder warp in tex	400	6 K
Carbon fibre density in g/cc	1.75	–

Table 16 Calculations (summarized) X, Y and Z contents—Noobed part (sample 4)

Content of warp (X)	
Content of stuffer warp (X1)	–
Total number of stuffer warp	0
Total length of stuffer warp in mtr	0
Weight of stuffer warp in gms	0
Content of Binder warp (X2)	–
Total number of binder warp	72
Total length of binder warp in mtr	14.49
Weight of binder warp in gms	5.80
Content of warp (X1) in gms	**5.80**
Content of Z threads	
Total number of binder warp	72
Number of Z threads for one binder thread	39.37
Total length of Z threads	7.13
Weight of Z threads in gms	2.85
Content of Z threads in gms	**2.85**

Table 17 Data obtained from preform—angle interlock part (sample 4)

Data		
Preform length in cm	20	–
Preform width in cm	23	–
Preform thickness in cm	**0.25**	–
Thread density	Per cm	Per inch
Stuffer warp (X1)	0.00	0
Weft (Y)	23.62	60
Binder warp (along width)—X2	18.90	48
Carbon fibre details		
Stuffer warp (X1) in tex	400	6 K
Weft (Y) in tex	400	6 K
Binder warp in Tex	400	6 K
Carbon fibre density in g/cc	1.75	–

Table 18 Calculations (summarized) X, Y and Z contents—angle interlock part (sample 4)

Content of warp (X)	
Content of stuffer warp (X1)	–
Total number of stuffer warp	0
Total length of stuffer warp in mtr	0.00
Weight of stuffer warp in gms	0.00
Content of binder warp (X2)	–
No. of pick columns on which binder thread repeats	**14**
No. of pick columns per inch	**9**
Linear distance occupied by 1 binder	3.95
Linear distance occupied by 1/2 binder	1.98
Half-length of binder thread	1.99
Length of one binder thread	3.98
Angle in degree	7.21
X2 component of 1/2 binder thread	1.98
X2 component of 1 binder thread	3.95
X2 component for 1 cm length preform	1.00
X2 component for total preform length	20.00
Total number of binder warp	435
Total length of binder warp in mtr	86.93
Weight of binder warp in gms	34.77
Content of warp (X) in gms	**34.77**
Content of weft (Y)	
Total number of wefts	425
Total length of weft in mtr	97.80
Weight of weft in gms	39.12

(continued)

Table 18 (continued)

Content of warp (X)	
Content of weft (Y) in gms	**39.12**
Content of Z threads	
Total number of binder warp	435
Number of Z threads for one binder thread	2.53
Total length of Z threads in mtr	11.00
Weight of Z threads in gms	4.40
Content of Z threads in gms	**4.40**

Table 19 Summarized fibre volume fraction for combined Noobed and angle interlock preform (sample 4)

Total preform volume in cc	115	
Total preform fibre content by volume	49.68	–
Total preform fibre volume fraction	**0.434**	**43.20%**
Volume of X fibres	23.18	–
Content of warp—X	**0.47**	**46.66%**
Volume of Y fibres	22.35	–
Content of weft—Y	**0.46**	**45.00%**
Volume of Z fibres	4.14	–
Content of Z thread	**0.08**	**8.34%**

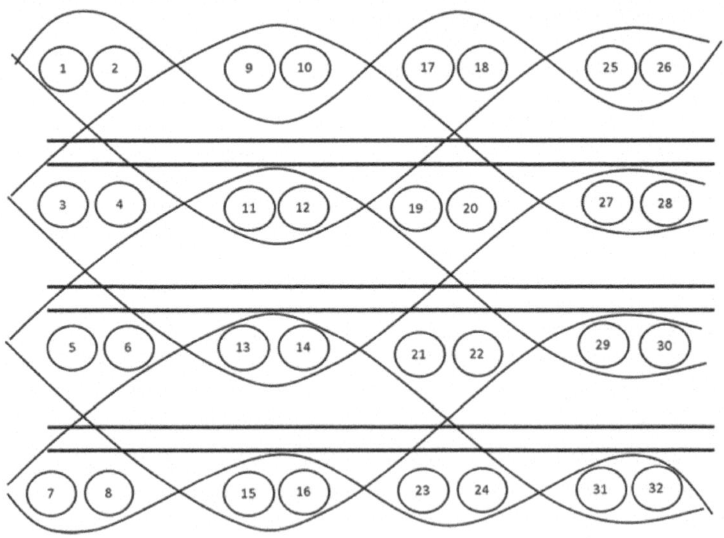

Fig. 13 Weave cross-section (weft way) of sample 5

No.	1	2	3	4	5	6	7	8	9	10	11	12	13	14	15	16	17	18	19	20	21	22	23	24	PPI	NOP	Beat
34	X	X	X	X	X	X	X	X	X	X	X														8	1	1
33						X																			8	0	1
32						X																			8	0	1
31	X	X				X		X																	8	0	1
30	X	X				X		X																	8	0	1
29	X	X	X			X	X	X		X															8	0	1
28	X	X	X			X	X	X		X															8	0	1
27	X	X	X	X	X	X	X	X	X		X														8	0	1
26	X	X	X	X	X	X	X	X	X		X														8	0	1
25	X	X	X	X		X	X	X	X	X	X														8	0	1
24	X	X	X	X		X	X	X	X	X	X														8	0	1
23	X	X	X			X	X	X	X																8	0	1
22	X	X	X			X	X	X	X																8	0	1
21	X	X				X	X																		8	0	1
20	X	X				X	X																		8	0	1
19	X																								8	0	1
18	X																								8	0	1
17	X	X	X	X	X	X	X	X	X	X	X														8	1	1
16							X																		8	0	1
15							X																		8	0	1
14	X	X					X		X																8	0	1
13	X	X					X		X																8	0	1
12	X	X	X			X	X		X		X														8	0	1
11	X	X	X			X	X		X		X														8	0	1
10	X	X	X	X	X	X	X	X	X		X														8	0	1
9	X	X	X	X	X	X	X	X	X		X														8	0	1
8	X	X	X	X		X	X	X	X	X	X														8	0	1
7	X	X	X	X		X	X	X	X	X	X														8	0	1
6	X	X	X			X	X	X	X																8	0	1
5	X	X	X			X	X	X	X																8	0	1
4	X	X				X	X																		8	0	1
3	X	X				X	X																		8	0	1
2	X																								8	0	1
1	X																								8	0	1

Fig. 14 Weave design as woven on the loom of sample 5

Fig. 15 Photograph of the preform (sample 5)

Table 20 Summarized data sheet for angle interlock layer to layer (Type 2)—sample 5

1. Preform code	AILLWL8PPC32PD6 8 weft layers, picks per cycle—32, pick density—6		
2. Raw material details a. Material b. Source c. Trade name d. Type (twisted or untwisted) e. Count f. Modulus	Carbon tows Tairyfil, Korea TC 33 Untwisted 400 tex Standard—230 GPa		
3. Preparatory process details Warping (beam or creel) Total No. of binder warp Total no. of stuffer warp	Creel 396 216		
4 Preform details	**As calculated**	**Actual obtained**	
Overall dimensions (L × B × H) in Cm	23 × 23× 0.3	22.5 × 22 × 0.25	
Total ends per inch for binder = No. of warp layers × surface ends/inch	8 × 4 = 32	8 × 4 = 32	
Stuffer threads per inch = No. of ends per dent × No. of stuffer dents/inch	6 × 4 = 24	6 × 4 = 24	
Total picks per inch = Picks per cycle × surface picks per inch	8X8 = 64	8 × 8 = 64	
Preform weight	98.6	100 + 10	Extra weave portion
Total fibre volume content	0.46		
X content (%)	46.25%		
Y content (%)	53.75%		
Z content (%)	NIL		
Weave cross-section and weave design as woven on the loom	Figs. 13, 14		
Summarized calculations of X, Y and Z content and fibre volume content	Tables 21, 22		
Sample photographs	Fig. 15		

Table 21 Data obtained from preform (sample 5)

Preform length in cm	22.5	
Preform width in cm	22	–
Preform thickness in cm	0.25	–
Thread density	Per Cm	Per inch
Weft (Y)	25.20	64
Binder warp (along width)—X	12.60	32
Stuffer warp (along width)	9.45	24
Carbon fibre details		
Binder warp (X) in tex	400	6 K
Weft (Y) in tex	400	6 K
Carbon fibre density in g/cc	1.75	–

Content of warp (X).
Crimped X_1:
8 Ends Per Dent in 4 Dents/inch (Alternate Dents) = 32 Ends/inch
Total number of binder ends for the whole width of fabric is 32*9 = 288 ends.
Length of each binder ends with 1% crimp is 225 + 2.25 = 227 mm = 0.227 m.
Total length of Binder yarn X1 is 0.227 × 288 = 65.37 m.

Non-crimped X_2:
6 Ends Per Dent in 4dents/inch (Alternate Dents) = 24 Ends/inch
Total number of stuffer ends in width of fabric is 24 × 9 = 216 ends.
Length of each stuffer end is 22.5 cm = 0.225 m.
Total length of Stuffer yarns X2 is 0.225 × 216 = 48.6 m.
Total Warp content is X1 + X2 = 48.6 + 65.37 = 114 m.
Total weight of Warp content is 45.6 g.

Weft Content.
No. of weft threads/ inch = 8 (No. of wefts per pick of insertion) X 8 (surface picks/ inch) = 64 picks/inch.
Total picks in 9″ length of fabric = 576 picks.
Length of each pick = 0.23 m.
Total length of picks = 132.48 m.
Weft weight = 53 g.

Table 22 Calculations (summarized) X, Y and Z contents and Fibre volume content for sample 5

Fibre volume content		
Total preform volume in cc	123.75	–
Total preform fibre content by volume	56.34	–
Total preform fibre volume fraction	**0.46**	**45.53%**
Volume of X fibres	26.06	–
Content of warp—X	**0.53**	**46.25%**
Volume of Y fibres	30.29	–
Content of weft—Y	**0.54**	**53.75%**
Total weight of the preform	**98.6 gms**	

6 Case Study of Weaving a Complex Geometrically Shaped Component (Sample 6)

A practical development was carried out by choosing a geometrically complex structure to understand the feasibility of utilizing the concepts of Noobing for shape weaving. The part had thickness variations, tapers and projections at two ends. The actual development was addressed step by step, wherein, first the thickness variations were addressed, then the tapers, subsequently projection at one end and finally the second projection and thus the complete preform. The development was done in three iterations. However, only the final iteration, which resulted in the final preform, has been discussed here. Figure 16 shows the isometric view of the part that was chosen to be woven using 3D Noobing .

To weave on a 2D loom, it is required to flatten the projections. Figure 17a, b shows the various sections and the dimensions of the flattened portions as is required to be woven on the loom. Different sections have been identified as well in the figure where there would be a change in weaving.

Broad outline of the weaving approach: The entire weaving of the preform was carried out using the Noobing approach. All the thicknesses, tapers and projections were woven on a 24 shaft dobby loom customized for weaving 3D fabrics. All the 24 shafts were used for weaving the shape. The variations of the thicknesses and projections were achieved weft way, hence the weave cross-section is warp way. Table 23 shows the distribution of the stuffer and binder threads used for obtaining the required thickness at various stations.

Referring to Fig. 17a portion C is the base preform for which the shuttle provided with the loom was used for weaving. Portion D used the Origami approach (discussed in Chap. 1) of weaving and had a separate shuttle. Portion E could be woven using two approaches viz., Double cloth approach wherein an additional Noobed preform emerges from the surface of the base noob preform at the respective stations. The other approach is using the pleat weaving concept. All the concepts are explained in Chap. 1. The development of this geometry calls for the use of multiple shuttles, simultaneous weaving of the base, projection 1 and projection 2. On completion of weaving and doffing from the loom, the projections are set upright and the Origami is opened up as explained in Chap. 1. It is to be understood that the description above is only a bird's eye view of the development as practical weaving would give more clarity on such developments. It is also required to understand that these types of preforms require 'Measured Weaving'. 'Measured Weaving' requires measurement of the different stations during weaving to start and stop the thickness changes, taper points and the projections so as to replicate the geometrical part in question. Sometimes, this calls for an iterative approach to match the dimensions. In the case of technical fibres such as carbon, there will be no elongation or crimp and therefore the iterations to arrive at the final preform will be less as compared to conventional textiles. Figure 18a–c (warp way) shows the weave cross-section for different portions of the preform. Figure 18d shows the alternate weave cross-section(approach 2) to

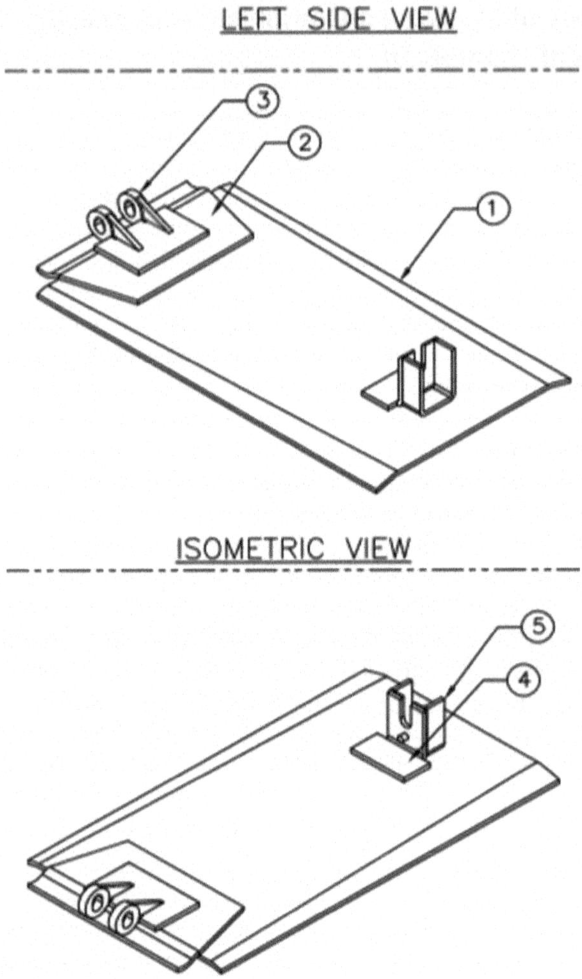

Fig. 16 Complex geometrical shape (sample 6). 1. Tapered edge 2. Zone of changed thickness. 3 Projection 1, 4. Zone of changed thickness 5. Projection 2

weave section E of the preform. This approach will also enable to double the thickness (as has been done in this case) to meet the component requirement. However, in the case of pleat weaving, even the weaving of the base is stopped to enable the formation of the pleat, and the reader is required to refer to Chaps. 1 and 2 for more details on pleat weaving. The weave designs and calculations are not discussed for this preform due to the complexity in understanding, interpreting, sequencing and relating it to the geometrical shape. Figure 19a–d shows the photographs of the preform (iterative mode) as it evolved stage wise. Figure 20a–h shows the photographs of the loom and the sub-elements used for weaving the samples.

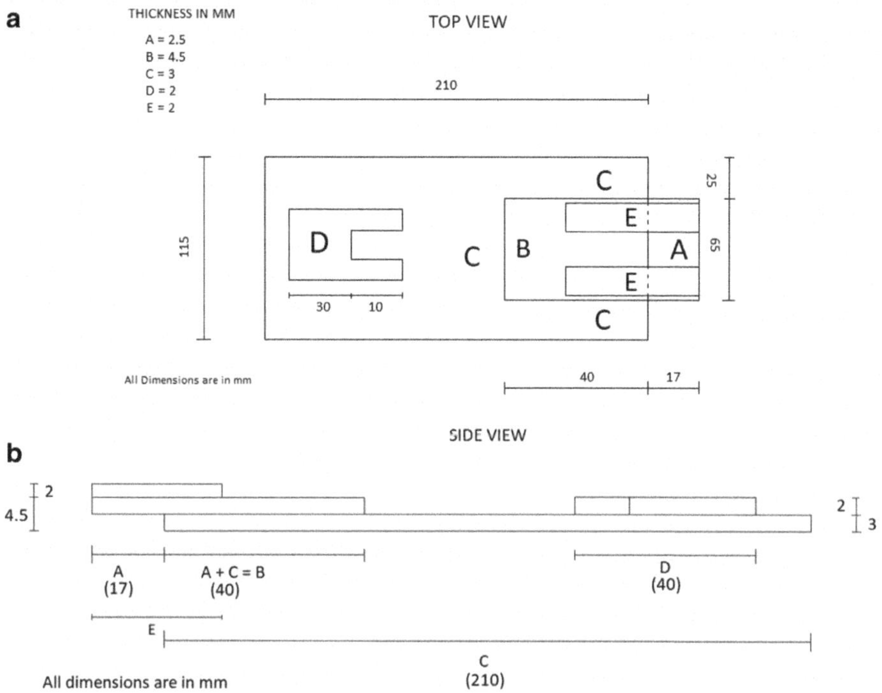

Fig. 17 **a** Top view of the flattened part wherein the projections are meshed onto the surface to enable weaving. **b** Side view of the different projections subsequent to flattening

Table 23 Distribution of stuffer and binder threads as per respective sections for sample 6

	A	B	C	D	E	Total
Stuffer warp	16	32	420	54	48	**507**
Binder warp	8	–	84	18	16	**126**

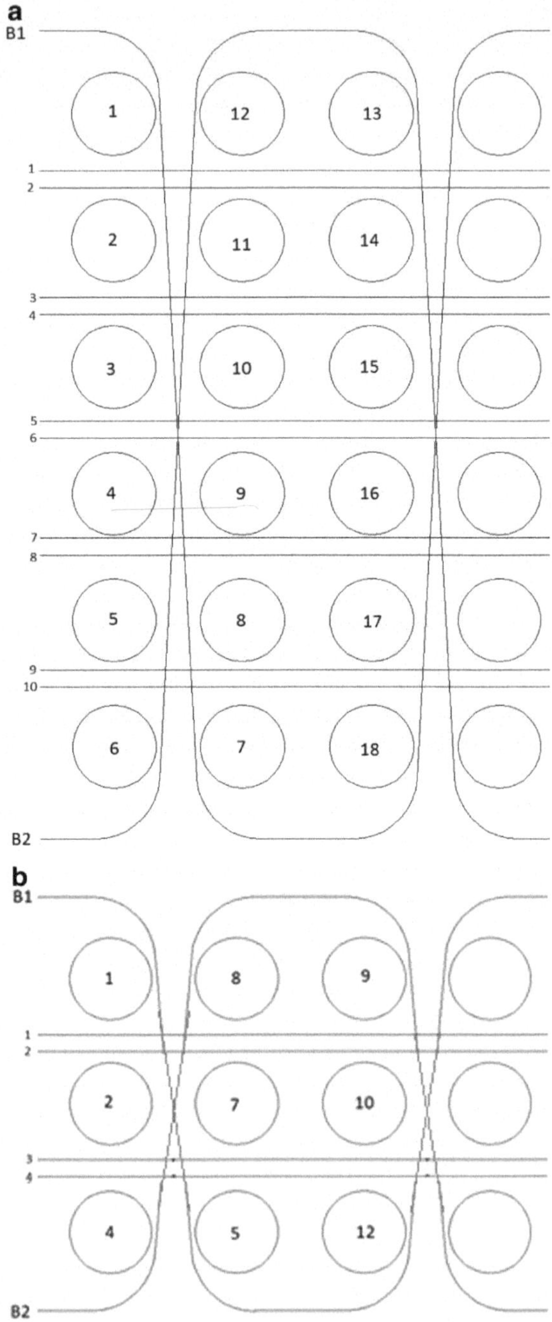

Fig. 18 **a** Weave cross-section (warp way) for section C of the preform (sample 6) base. **b** Weave cross-section (warp way) for sections A, B and D of the preform (sample 6). **c** Weave cross-section (Warp way) for section E of the preform (sample 6)—Approach 1. **d** Weave cross-section (Warp way) for section E of the preform (sample 6)—Approach 2

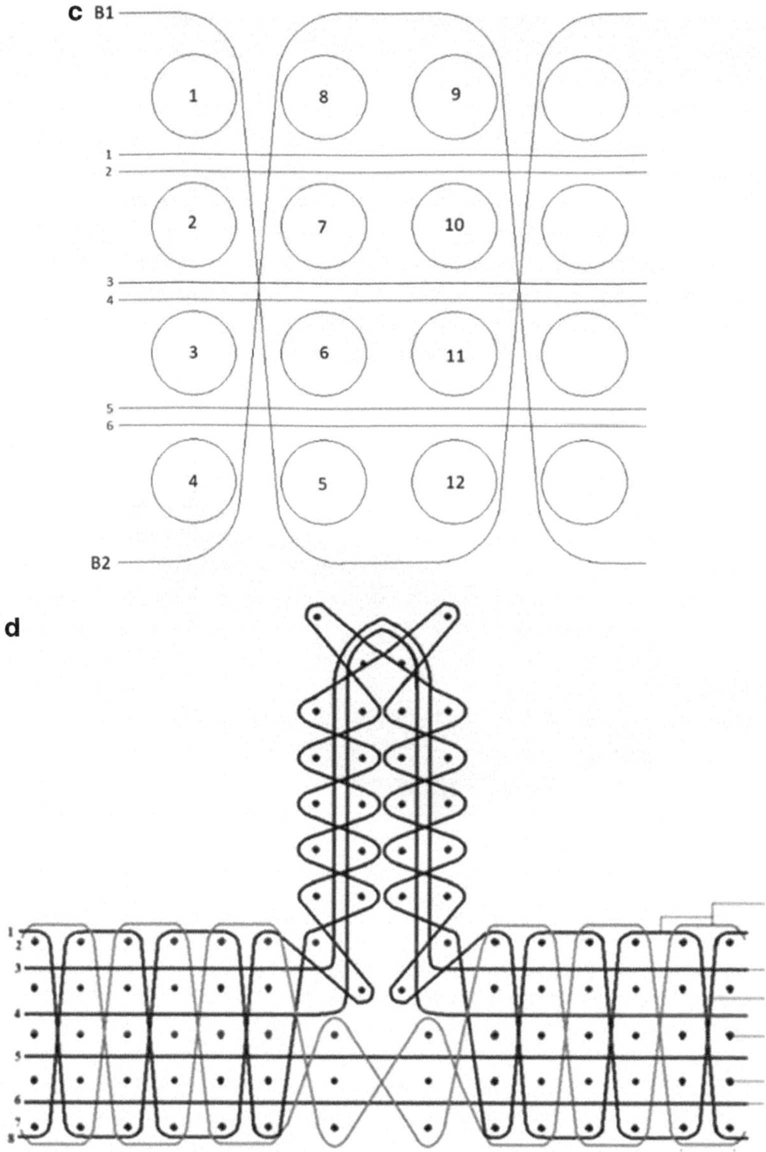

Fig. 18 (continued)

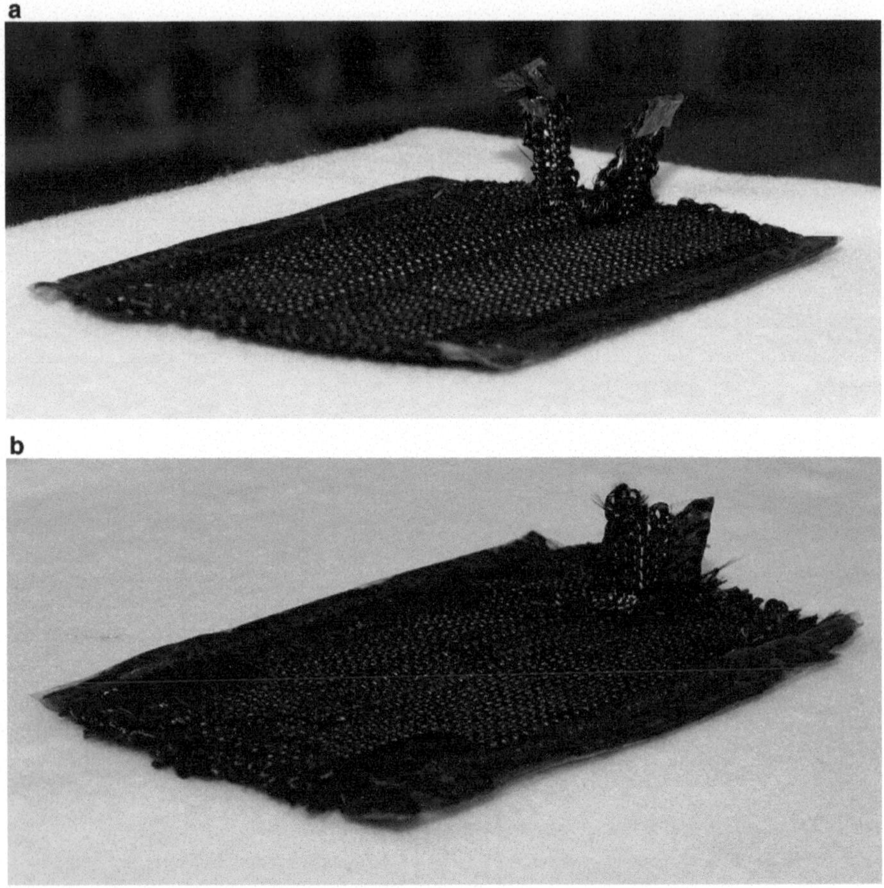

Fig. 19 **a** Stage 1 development (sample 6)—Thickness changes and projection 1. **b** Stage 2 development (sample 6)—Thickness changes and projection 2 based on the Origami Principle. **c** Stage 3 development (sample 6)—Final integrated preform. **d** Stage 4 development (sample 6)—Final integrated preform with thickness changes, projection 1 (using pleat weaving approach) and projection 2 (Origami based)

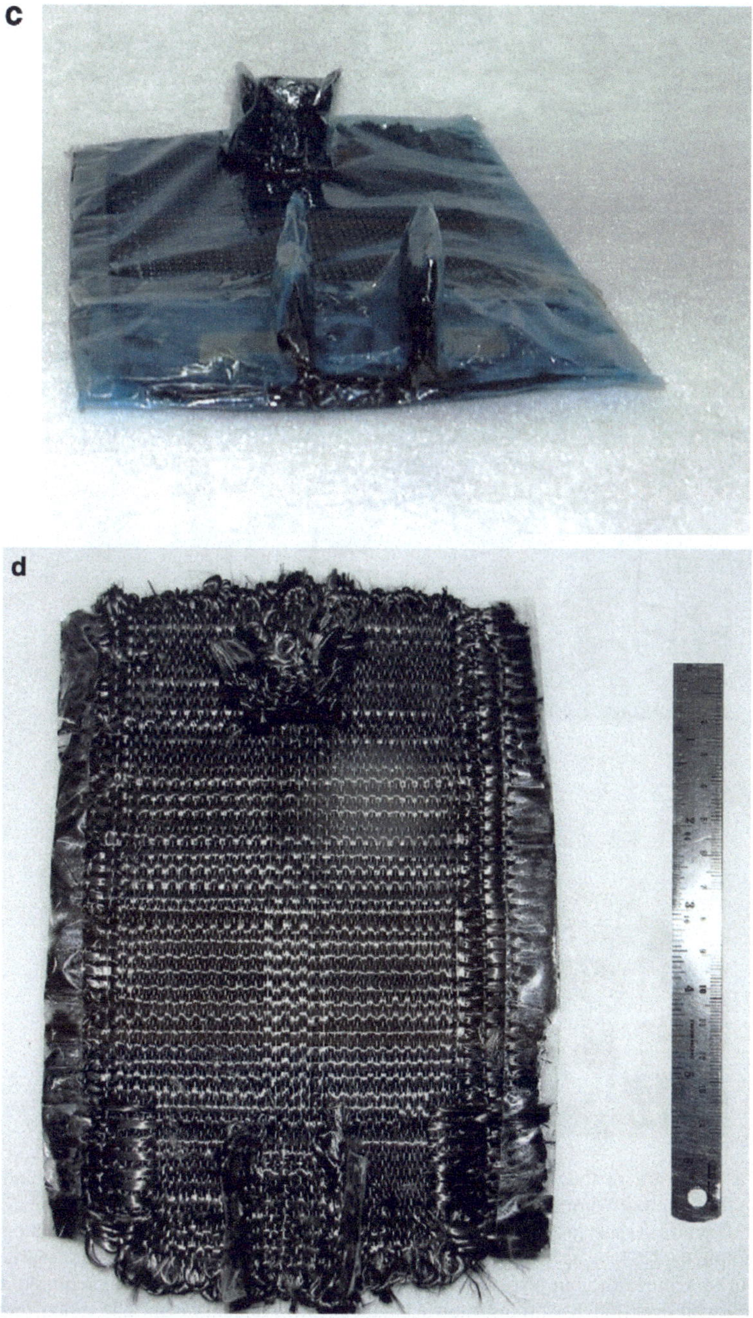

Fig. 19 (continued)

a

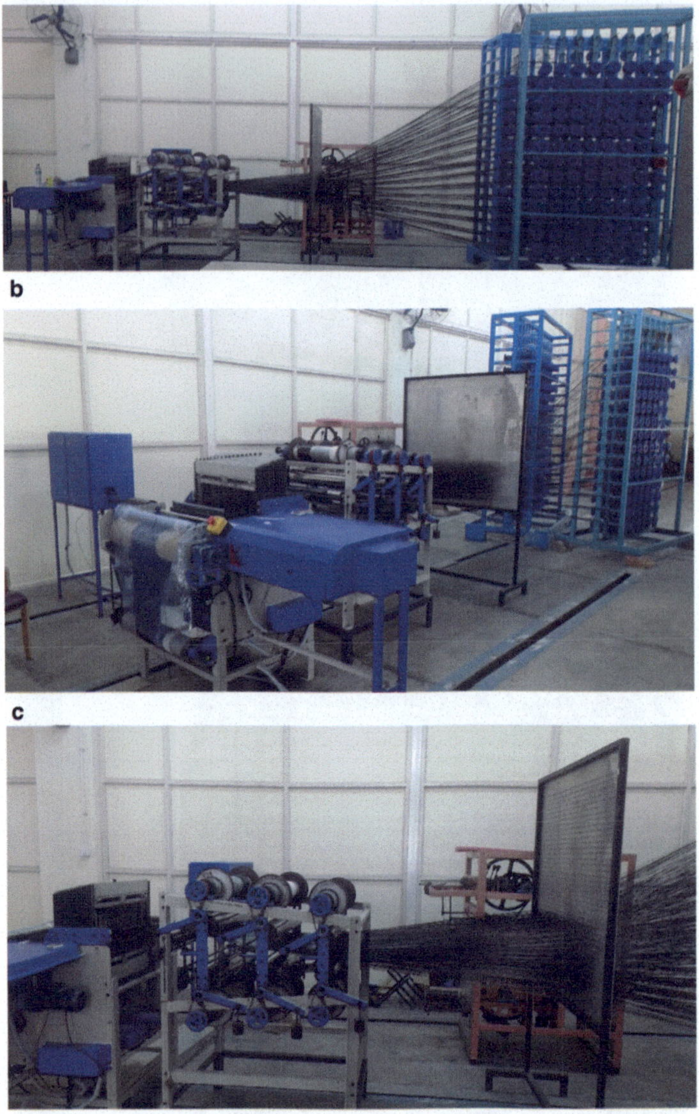

Fig. 20 **a** Side view of the 3D weaving loom. **b** 3D weaving facility (for Noobing and Angle Interlock) front view. **c** Warp sheet through the perforated sheet and beam stand entering the loom. **d** Woven 3D preform sheet. **e** A portion of the software design for weaving the complex geometrical shape (sample 6). **f** Close-up view of the varied drawing-in pattern of the 24 heald shafts during drawing-in to achieve the out-of-plane projections and thickness variations as required. **g** Shed opening for weft insertion. **h** Weft insertion using the positive rapier system. The yellow threads on the reed display the different stations identified for drawing in the warp (varied content) to account for the thickness changes and projections.

d

e

Fig. 20 (continued)

Fig. 20 (continued)

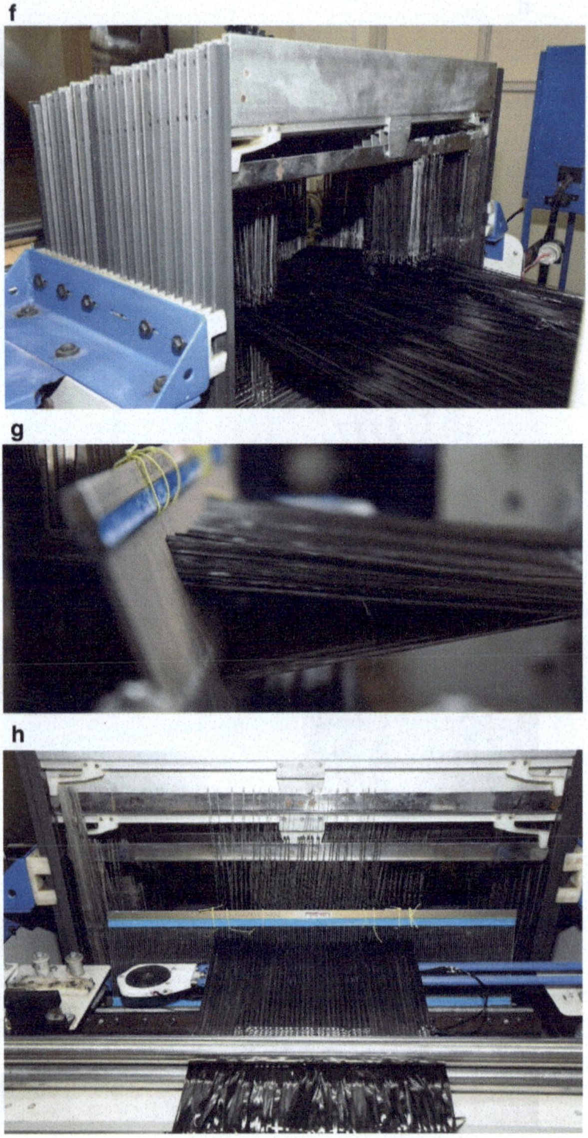

Chapter 4
Dual Plane Shedding Based 3D Weaving

B. S. Sugun, J. Ramaswamy Setty, Shashikant Choudhary, and G. N. Dayananda

1 Introduction and Basics

Three dimensional weaving technology based on dual plane shedding is a specialized technology for the development of interlaced 3D reinforcements utilizing the textile weaving process. Applications for this technology exist in composites used in aerospace, automobile, shipbuilding, construction, medical, body armour industries and the like.

In its ideal sense, dual-plane shedding-based preforms are akin to 2D woven preforms with interlacements and related paraphernalia and are considered by few as "True 3D Weaving". The technology is pursued by very few researchers in the world due to the following reasons.

1. Machine design and development is complex and totally deviates from the conventional weaving loom design hitherto existing in the textile industry
2. The technology is limited to "stand alone Structures" in composites, due to the cross-sectional limitations, most importantly, where damage tolerance becomes an over-riding criteria

Weaving based on dual plane shedding, is about weaving multiple layers of cloth in the X, Y and X, Z planes as well as interweaving them together to form block

B. S. Sugun (✉) · J. Ramaswamy Setty · S. Choudhary · G. N. Dayananda
Centre for Societal Missions and Special Technologies (CSMST), CSIR-National Aerospace Laboratory, Bangalore, India
e-mail: sugun@nal.res.in

J. Ramaswamy Setty
e-mail: ramaswamy@nal.res.in

S. Choudhary
e-mail: Shashi_nal@nal.res.in

G. N. Dayananda
e-mail: dayananda@csir.res.in

B. S. Sugun (ed.), *Practical Approach to 3D Weaving*, Materials Horizons: From Nature to Nanomaterials, https://doi.org/10.1007/978-3-030-80424-4_4

or profiled preforms. The technology is in its nascent stage of development around the world. The specialities of dual-plane shedding-based 3D weaving as compared to the more popular orthogonal weaving and angle interlock weaving are as follows.

1. The X, Y and Z threads are all orthogonal and interlaced similar to 2D woven preforms thus making it a solid interlaced 3D structure
2. In the case of profile weaving, yarn architecture will be intact and as per the desired profile, the warp threads will be positioned as per the specific profile and vertical and horizontal wefts are woven around that

The complexity associated with the technology is that, the machine development is entirely from the first principles of textile loom construction, wherein, each of the sub-elements is required to be designed and developed using novel principles and is in fact, deviating from the conventional approach of loom design. Although there are arguments that the crimp introduced due to the interlaced structure will affect the in-plane strength, the trade-off would be appreciable, as out-of-plane strength increases, which in turn improves the damage tolerance of the structure.

As is the case with any technology, the identified drawbacks of dual-plane shedding-based 3D weaving are as follows.

1. Complexity in loom design, manufacture and therefore, pursued by very few in the world
2. Damage caused to yarns particularly to technical textile yarns such as carbon and glass due to the interweaves
3. Preform cross-section is limited. As per the current feasibility study carried out, a maximum preform cross-section of 250 $\times$ 250 mm is possible, although the current world standard as found in literature is 60 $\times$ 60 mm

However, in the authors' opinion, this is a technology worth pursuing by the research community for modular joints as well as for the "stand alone Structures" in composites, such as the links and fitting of "wing-fuselage joints", wherein, there would be a significant weight reduction, coupled with improved performance.

This chapter focuses to understand the technical intricacies and associated solutions involved in the development of the technology.

2 Working Principle Description

Dual-plane shedding-based three dimensional weaving as such is a complex technology requiring the formation of sheds in two planes, picking in those planes and suitable let-off, beat-up and take-up mechanisms. Although, handlooms exist even to this day in the 2D weaving industry (to the extent that some intricate designs are woven by hand), 3D weaving on a non-mechanized loom would be impossible. The main reason for the requirement of mechanization is that the technology demands the following:

1. Simultaneous formation of sheds in all the layers.
2. Simultaneous insertion of two or more shuttles (as many as the number of warp layers) in all the sheds of any given plane.
3. Simultaneous beat-up of all the inserted wefts in any given plane.

In addition to the above, sequential operation as described in the weaving cycles would be required, which calls for linking and timing the operations appropriately. In addition to the above, mechanization also ensures quality viz., uniformity and repeatability in the developed 3D structures.

For conventional weaving, the warp is arranged in the form of sheets. Even in the case of angle interlock weaving and noobing, warp arrangement will be in the form of several collapsed layers of sheets one above the other. However, dual plane shedding calls for an entirely different type of arrangement of warp which is required to be in the form of a grid. To understand the requirement of the grid, it would be required to first understand the weaving concept of dual-plane shedding-based 3D preform. Figure 1a–h shows the same and is explained step-by-step below.

- Figure 1a represents the warp arrangement in grid form (termed warp in home position). Grid arrangement is preferred to enable warp movement for shedding in both the planes viz., horizontal and vertical planes
- Figure 1b represents the vertical movement of alternate(odd) columns of warp to create space for horizontal weft insertion
- Figure 1c represents simultaneous horizontal weft insertion in all the layers of the separated warp
- Figure 1d represents the warp in the home position
- Figure 1e represents the horizontal movement of alternate rows of warp(odd) to create space for vertical weft insertion
- Figure 1f represents simultaneous vertical weft insertion in all the layers of the separated warp
- Figure 1g represents the warp in the home position. With this, one half of the 3D weaving cycle is complete.
- Figure 1h represents the vertical movement of alternate(even) columns of warp to create space for horizontal weft insertion
- Figure 1i represents simultaneous horizontal weft insertion in all the layers of the separated warp
- Figure 1j represents the warp in the home position
- Figure 1k represents the horizontal movement of alternate rows of warp(even) to create space for vertical weft insertion
- Figure 1l represents simultaneous vertical weft insertion in all the layers of the separated warp
- Figure 1m shows the warp in the home position. With this the 3D weaving cycle is complete
- Figure 1n shows the cross-section of the 3D dual-plane shedding-based woven preform. One unit cell of 3D woven preform is complete and based on the length required weaving is carried out as detailed above on a continuous basis.

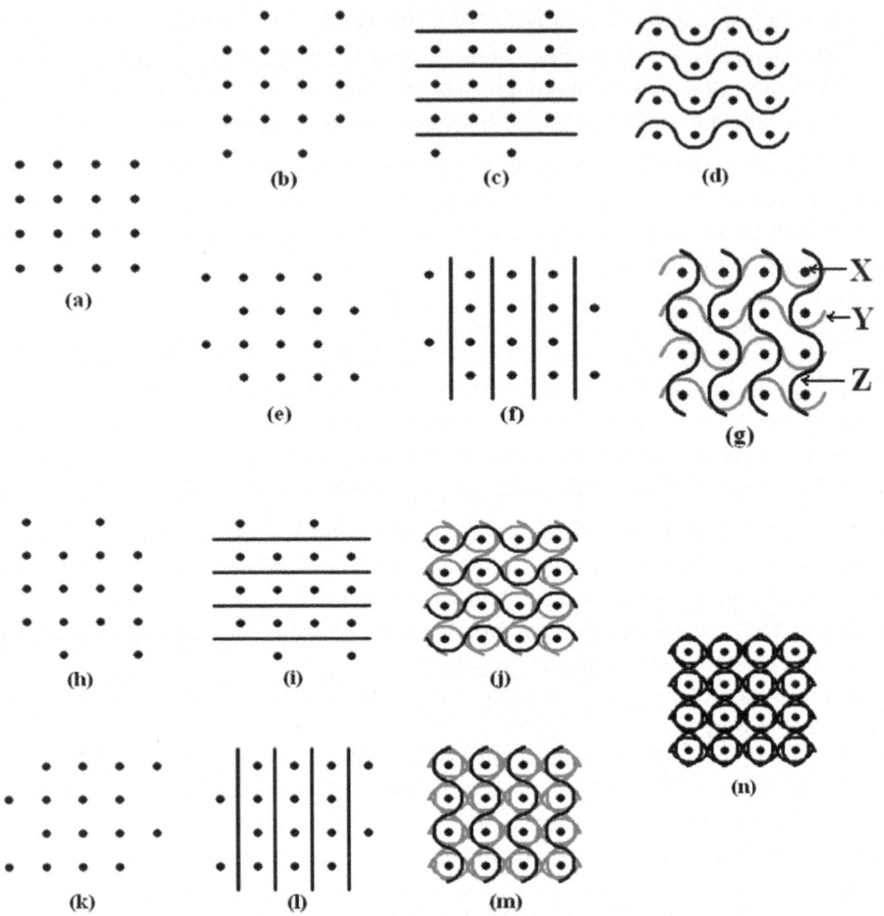

Fig. 1 3D dual-direction shedding weaving approach. **a** Grid like warp (X) arrangement (Home position). **b** Vertical shedding. **c** Horizontal weft (Y) insertion. **d** Home position. **e** Horizontal shedding. **f** Vertical weft (Z) insertion. **g** Home position and completion of half weaving cycle. **h** Vertical shedding. **i** Horizontal weft insertion. **j** Home position. **k** Horizontal shedding. **l** Vertical weft insertion. **m** Home position and completion of full weaving cycle. **n** 3D woven preform

3 Previous Works Related to Dual Plane Shedding Based 3D Weaving

Dual plane/direction shedding-based 3D weaving has been pioneered by Biteam, wherein, they are utilizing the technology to develop several complex profiles. The principle of working is elucidated in [1, 2] and Fig. 1. Experimental studies [3] using vinyl ester resin-based composites have shown higher out-of-plane tensile strength and shear strength with changes in the failure modes compared to the traditional 2D laminates.

A plethora of information available in literature domain (especially patents) discusses about various approaches to develop 3D fabrics. US patent Nos. 3834424, 3904464, 3955602, 4019540, 4526026, 5137058, 5767023, 5772821/5833802 and 2014/0360618 A1 describe machines that produce non-interlaced 3D fabrics. US patent No. 6186185 B1/6431222 B1, 6338367 B1, 1512/CHE/2005 describes the dual plane shedding approaches, and WO 01/66840 A2 describes weft insertion and beat-up in two planes.

4 Theoretical Information Governing Machine Design

The requirements of three dimensional weaving is similar to that of two-dimensional weaving, in that, it needs let-off mechanism for the warp yarns, shedding mechanisms, picking mechanisms, beat-up and take-up mechanisms. However, the approach will be entirely different when compared to the 2D weaving process. Figures 2, 3 and 4 show the 2D and 3D weaving cycles. These weaving cycles give a fair idea of the process as well as the complexity involved in developing 3D woven reinforcements as compared to 2D fabrics. In Figs. 3 and 4 there is a possibility of merging the primary and secondary beat-up based on the design of the weaving machine. Figure 5 shows the difference in utilizing the two weaving cycles to weave the T preform. Figure 6 shows the possible profiles (not limiting) that can be woven using dual-plane shedding-based 3D weaving.

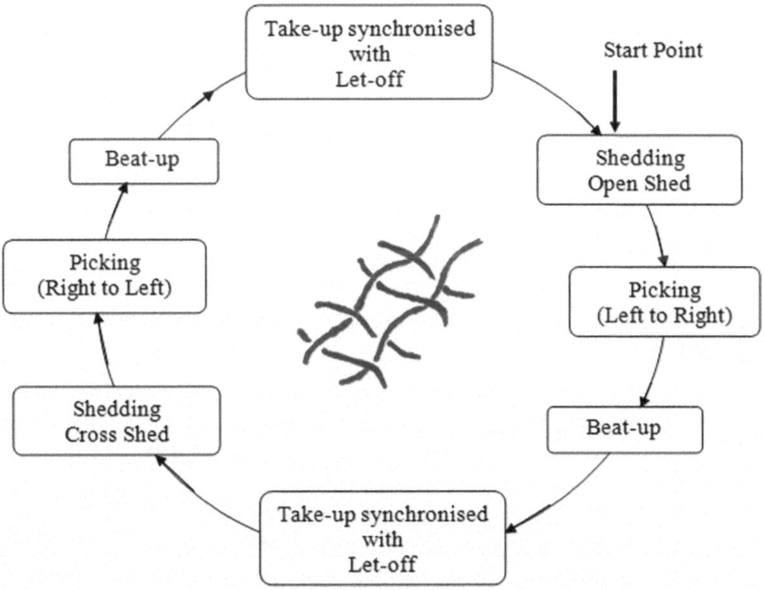

Fig. 2 Weaving cycle for 2D weaving as it happens step-wise—(Inset–Plain Woven Architecture)

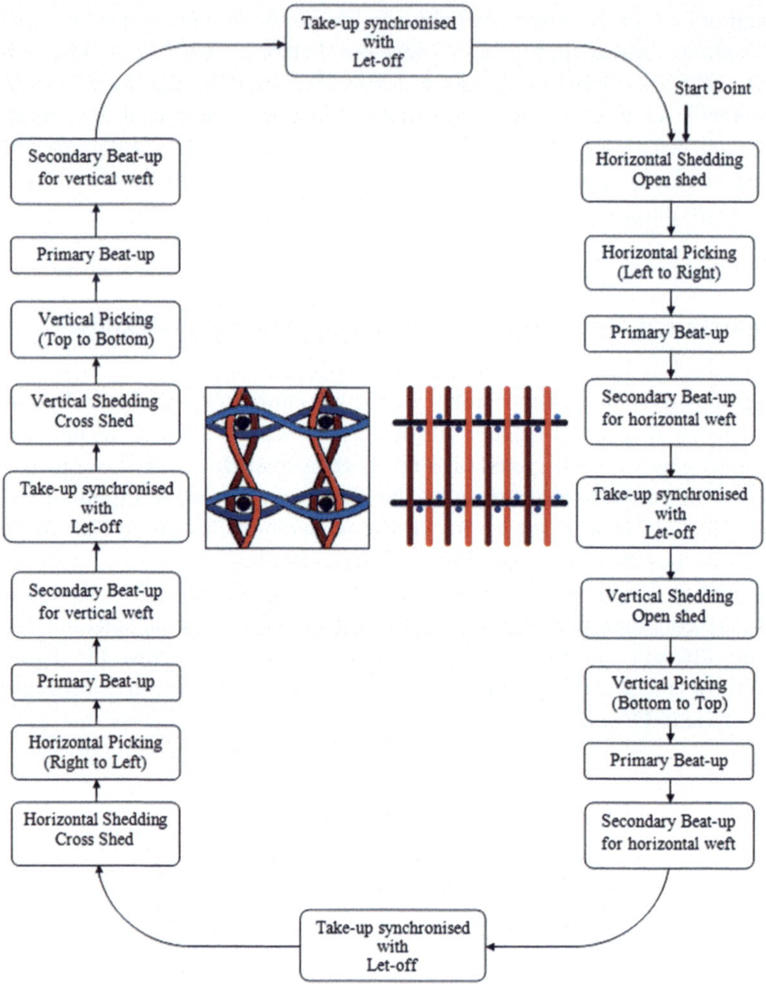

Fig. 3 Weaving cycle-1 for 3D weaving as it happens step-wise for block structures (Ideal)—(Inset–3D Plain Woven Architecture)

4.1 Shedding

The warp for 3D weaving is required in several layers. The conventional beam as used in 2D weaving, cannot accommodate this, nor would it be practical to have beams for each layer. Therefore, warp let off for 3D weaving would be practical if the creel system is used. Since creel is of common occurrence in the textile industry, the mechanization details are not elaborated here. However, to specifically adopt the creel to 3D weaving of technical textiles such as carbon preforms (which is a

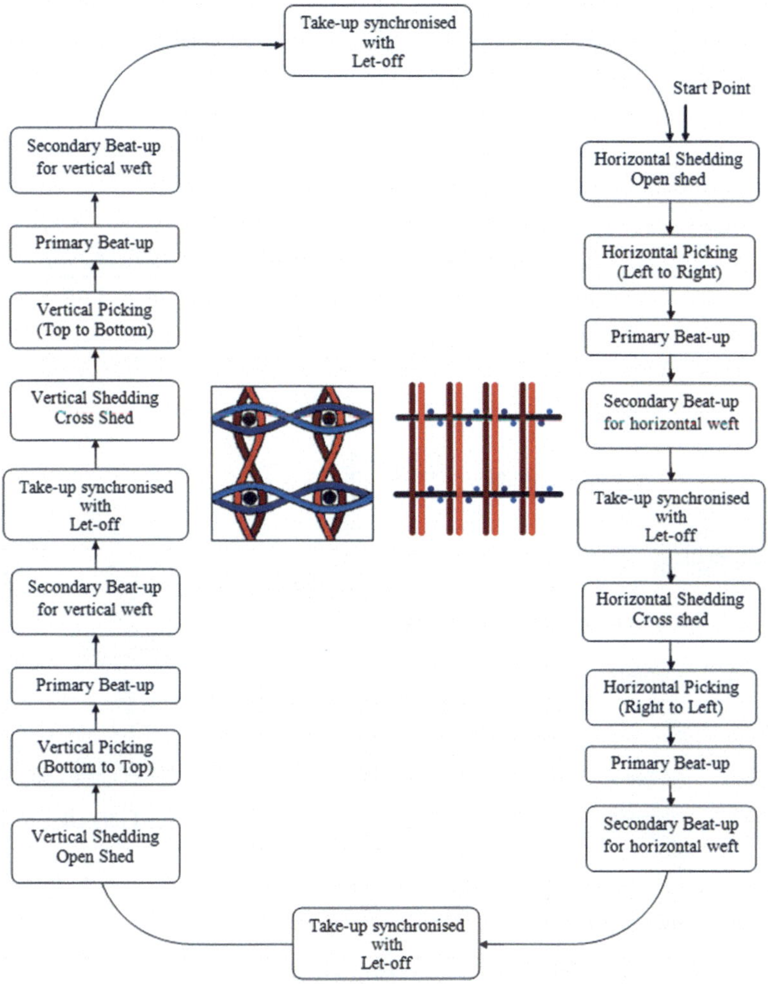

Fig. 4 Weaving cycle-2 for 3D weaving as it happens step-wise for profiles—(Inset–3D Woven Architecture)

strong contender for 3D technology), tensioning devices are modified. An indirect tensioning system is adopted to prevent damage to carbon.

The commercially available creels have spring/dead weight type tensioners, wherein, the yarn passes between two well-polished plates and the tension is directly applied on the yarns by means of the spring force or the weight of the dead weight. Here, an indirect tensioning device is adopted to prevent abrasion of the carbon yarns. The yarn package itself is tensioned by having an additional roller attached to the double-flanged bobbin over which a rubber belt runs. One end of the rubber belt is fixed to the roller, while the other end has a free-hanging hook which can be weighed down by the addition of washers. Increasing/decreasing the number of

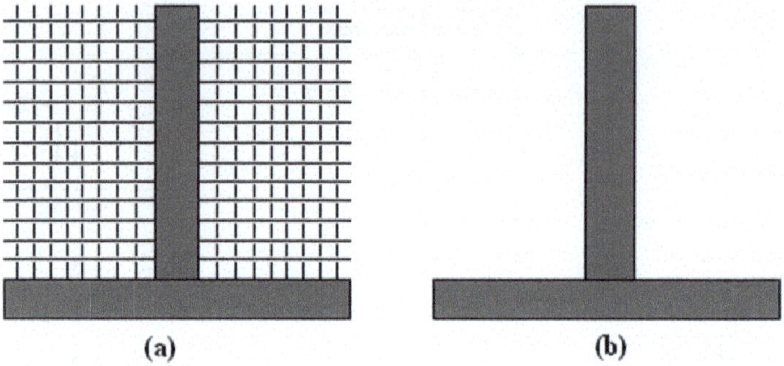

Fig. 5 **a** Structure of T woven using weaving cycle-1. **b** Structure of T woven using weaving cycle-2

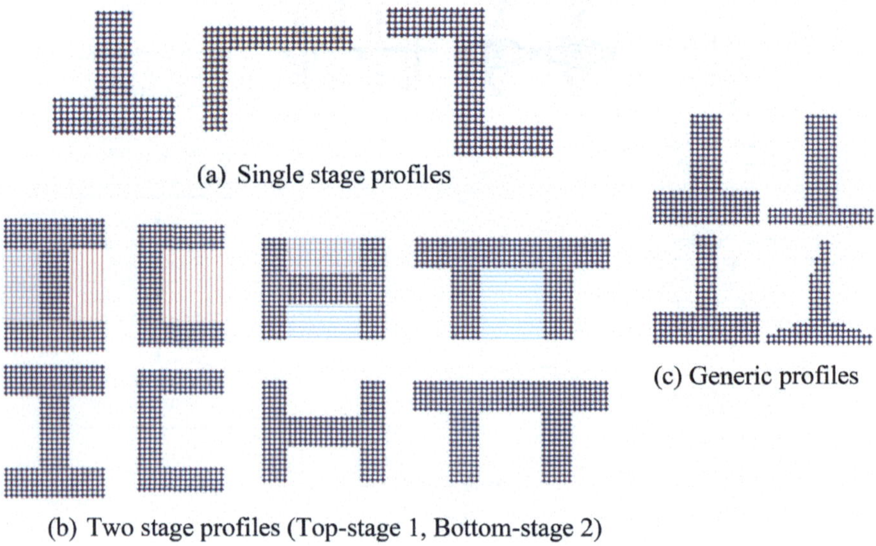

Fig. 6 Possible profiles that can be woven using dual-plane shedding-based 3D weaving

washers increases or decreases the tension in the warp yarns. Figure 7 shows the line sketch of the tensioning device for let-off for carbon tows.

Warp congregate sheet placed between the creel and the machine comprises of equally spaced holes arranged in a grid manner bonded with ceramic eyelets(for carbon passage). This will aid the arrangement of warp in grid-like manner and enables passage of the warp threads into the weaving machine.

The warp congregate sheet is a thin plate of about 3 mm thick stationed behind the shedding zone. Although it is not the main element, it is required for the channelization of the warp yarns. It has holes arranged in grid form with the spacing of the warp threads throughout. The number of holes in the warp congregate sheet will be equal

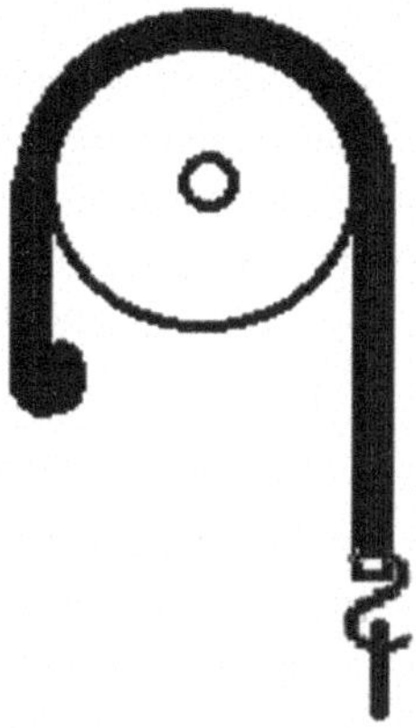

Fig. 7 Line sketch of the tensioning device for the let-off

to the total number of warp yarn. Each hole will be bonded with a ceramic eyelet to aid the passage of the yarn without abrasion. Each warp thread passes through one hole. The purpose of this sheet is twofold. One, it aligns all the warp threads coming from the creel in such a manner that they are all fed parallel to the shedding assembly which is required for shed formation. Two, it maintains the designated spacing between individual warp yarns and also helps in locating its position in case there is a breakage of thread or the particular warp bobbin gets exhausted.

The primary objective of the shedding device is to separate the warp threads to enable the weft insertion system to pass through. This is called shed opening. In the case of 3D weaving, it has to be done in two planes as shown in the schematic of Fig. 1.

In 2D weaving, the warp for the shed separates in only one direction. Therefore, the threads can simply lie together one adjacent to the other. In the case of dual-plane shedding-based 3D weaving, the warp needs to be separated in both directions alternatively. The spacing of the threads is a crucial parameter as it decides the shed depth and the location of the beat-up point. If the warp yarns are too close, there would not be enough space for the weft insertion device to pass through but would be preferred for beat-up and preform compaction. If the warp yarns are too far apart, beat-up mechanism would get complicated by having to be moved for a large distance and therefore preform compaction would get affected. The spacing is also a function of the size of the mechanism/system that is capable of inserting the weft in the least possible gap/space between the warps.

Now if the threads are to lie together in both the planes, then the shedding system design becomes complex, in that, during shedding each layer needs to be separated in the respective planes. Hence the arrangement of the warp itself will be made with some spacing between each thread which results in a grid-like arrangement. However, the preform at the fell, on compaction, will have the horizontal and vertical warp together after weaving. This, therefore, results in a natural convergence of the warp threads. Now, the convergence distance depends on the spacing of the warp in the grid. The spacing of the warp in the grid depends on the design of the weft insertion system. It is to be understood that unlike 2D weaving, there is no liberty for thick shuttles. Thinner shuttles will be the preferred route. This is because thick shuttles would need large shed space to pass through and therefore would result in increasing distance between the individual threads. This would result in increasing converging angle and in turn, increase in the beat-up distance. The thinner shuttle would minimize both the converging angle and the beat-up distance.

The central point of weaving loom design is the shed geometry as shown in Fig. 8 below. A typical shed geometry having the basic details is shown below (representative only). Three variants of the shed geometry can be seen in Figs. 8, 9 and 10. The figures are self-explanatory. What is required to be noticed is that, the central requirement would be to increase the shed space at the picking plane. However, a mere increase of 6 mm at the picking plane results in a shift of about 400 mm for the whole grid at the shedding zone. Figure 11 shows the practical trial of a layer of converging warp and weaving trial as per geometry. Figure 12 shows the photograph of warp yarns converging from the shedding zone(right) to the beat-up zone(left). In essence, the shed space should be based on the thickness of the shuttle design. It should be noted that picking happens simultaneously in all the layers and therefore a space of 1.5 mm both at the top and bottom of the shuttle in each layer is mandatory, which makes it about 3 mm between shuttles. Therefore a shedding space

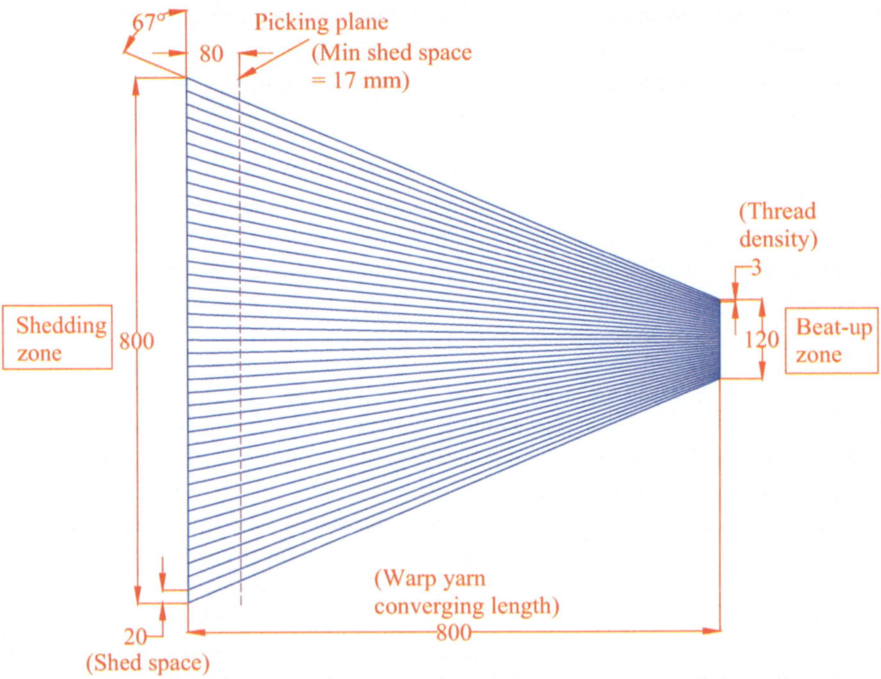

Fig. 8 Shed geometry for shed space of 17 mm at picking plane. Converging angle at the extremities—67°, shed space at picking plane—17 mm, picking length—800 mm

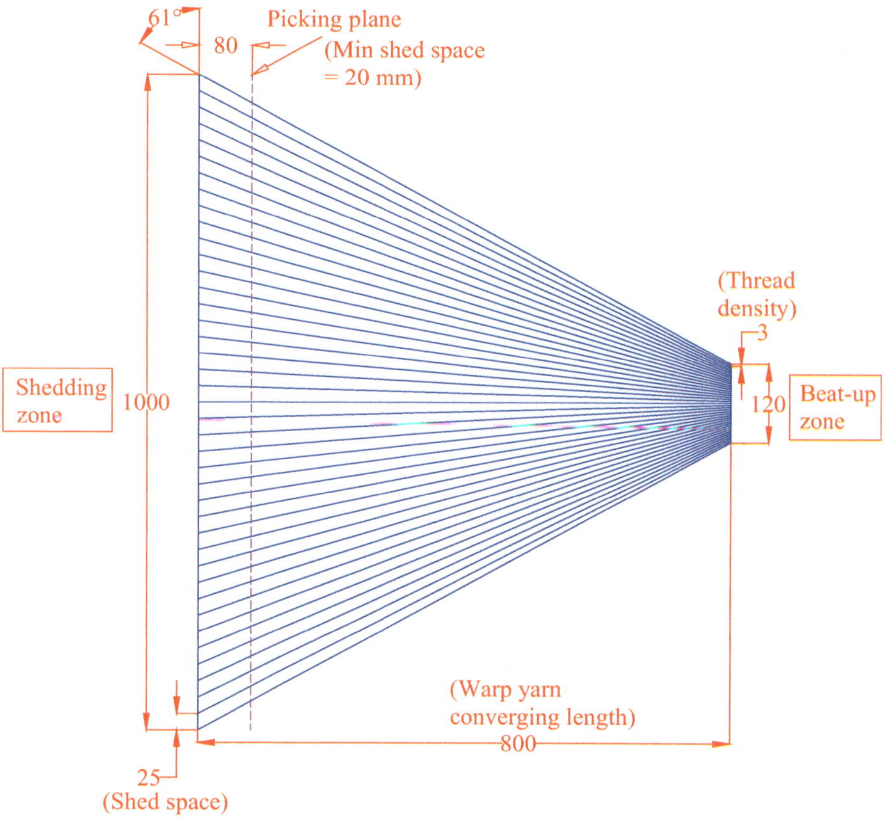

Fig. 9 Shed geometry for shed space of 20 mm at picking plane. Converging angle at the extremities—61°, shed space at picking plane—20 mm, picking length—1000 mm

of 20 mm would invariably restrict the shuttle size to less than 17 mm. Now with the shuttle size of just 17 mm, the entire technology of the weft insertion device needs to be designed. While it may be argued that there are slimmer weft insertion devices such as projectiles, it is to be understood that they are mere weft transporters and not weft carriers. It is to be understood that for dual-plane weaving a closed selvedge would be ideal as shuttleless weft insertion will add the additional complexity of locking the weft. This is not practical given the large number of layers as well as the two planes.

4.2 Picking

There are three major requirements in the picking process of 3D weaving. The first requirement would be the need for positive picking, wherein, the picking device needs to be always held against the negative picking in the shuttle looms or projectile

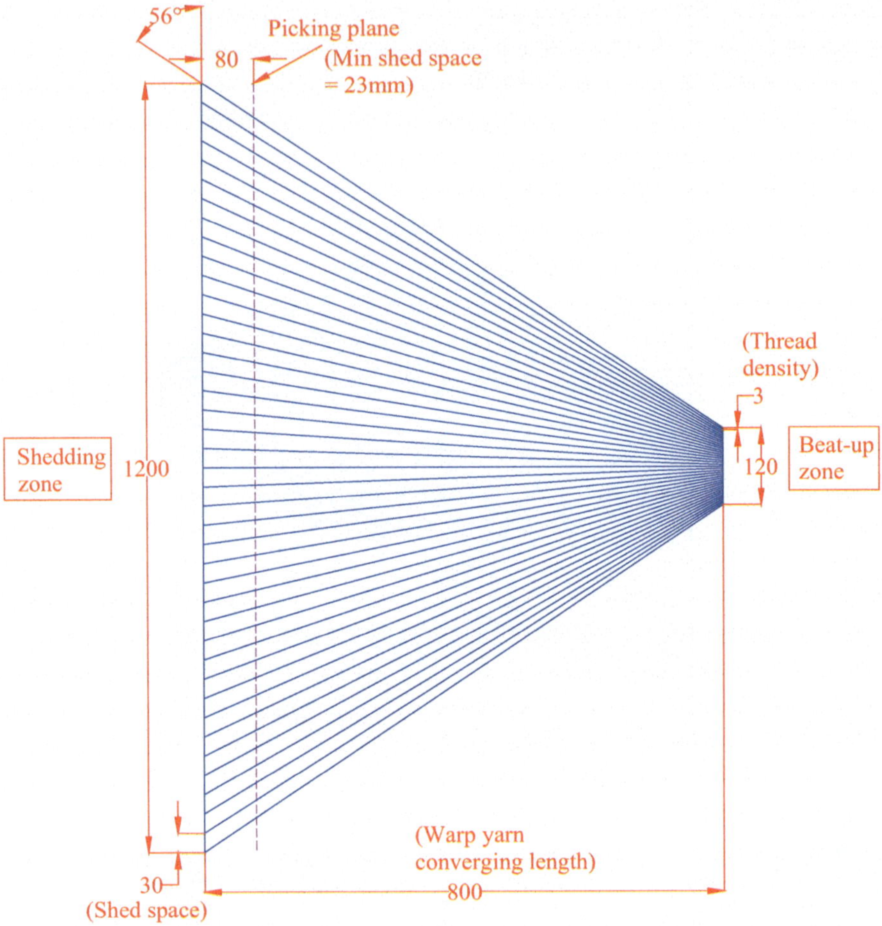

Fig. 10 Shed geometry for shed space of 23 mm at picking plane. Converging angle at the extremities—56° shed space at picking plane—23 mm picking length—1200 mm

looms of a 2D weaving machine. The concept of a race board or guiding device is not practical as it cannot be implemented for all the layers in 3D weaving. Also, the shed space in each layer is relatively small since 3D weaving has many layers in any given plane and therefore, it has to accommodate more number of sheds. The above argument therefore calls for a radically different shuttle design emphasizing on thinner and lighter ones. The second major requirement of the picking system, would be, provision for the weft insertion even in the vertical plane, which is obvious for adding reinforcement in the third direction. So, the machine would have two sets of picking assemblies (comprising of several picking devices and connecting links—one for each layer), drive mechanisms one for each of the horizontal and vertical weft insertions. The third major requirement of the picking process would be the

Fig. 11 Practical trial of a layer of converging warp and weaving trial as per geometry

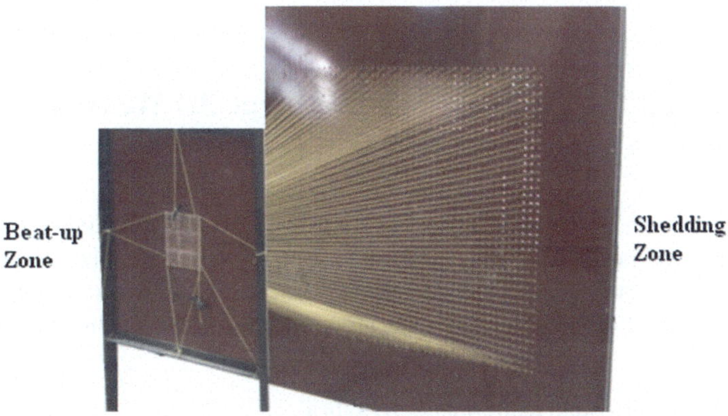

Fig. 12 Photograph of warp yarns converging from the shedding zone(right) to the beat-up zone(left)

requirement of the shuttles to be oriented in accordance with the particular angle of the shed at which it is stationed. This requirement would be due to the convergence of the warp yarns which obviously take different angles from the centre to the extremities in both the planes.

In addition to the above requirements, there would be the need for the mechanism to simultaneously insert wefts in all the layers of the structure both during horizontal and vertical picking. This would be a pre-requisite because the shed (open or cross) for all the layers in any given plane happens simultaneously and the pick insertion by positive means would be a relatively slow process and therefore simultaneous insertion ensures comparatively faster preform formation.

4.3 Beat-Up

The distance between the shedding zone and beat-up zone in 3D weaving is relatively larger when compared with the 2D weaving process. Moreover, the reed employed in the 2D weaving process cannot be adopted here due to the presence of grid-like warp yarns, as well as, picks in both horizontal and vertical planes. Here, dual beat-up (termed primary and secondary) would be required for preform compaction.

4.4 Take-Up

The conventional take-up on a 2D weaving loom uses rollers to wind the cloth. 3D woven structures will be thick and in profile form will have a complex shape wherein use of roller for winding is not possible. The preform has to be taken up in linear form.

5 Design and Development of a Prototype 3D Weaving Machine

Based on the above concepts, CSIR–NAL has designed and developed a prototype 3D weaving machine. The loom was designed from the first principles. The complete description of its construction and working is described below along with relevant line sketches and photographs. The governing criteria for the machine design are as below.

1. To obtain orthogonally interlaced 3D fabric.
2. To achieve dual plane shedding for 3D weaving of blocks and profiles.
3. To achieve the 3D weaving with the minimum required spacing of the warp yarns
4. To achieve the above 3D weaving with the inclusion of non-interlacing (stuffer) yarns

The 3D weaving machine uses the concept of a counter rotating tube and shaft of shedding assembly for obtaining vertical shedding required for horizontal weft insertion and translational movement of successive shedding assemblies in opposite directions for obtaining horizontal shedding required for vertical weft insertion.

The design of the 3D weaving machine involves the following.

1. Providing congregate sheet for guiding of yarns.
2. Design of mechanism to actuate multiple counter rotating tube and shaft assembly for dual plane shedding
3. Design of lightweight coupled with high stiffness, special shuttle for weft insertion

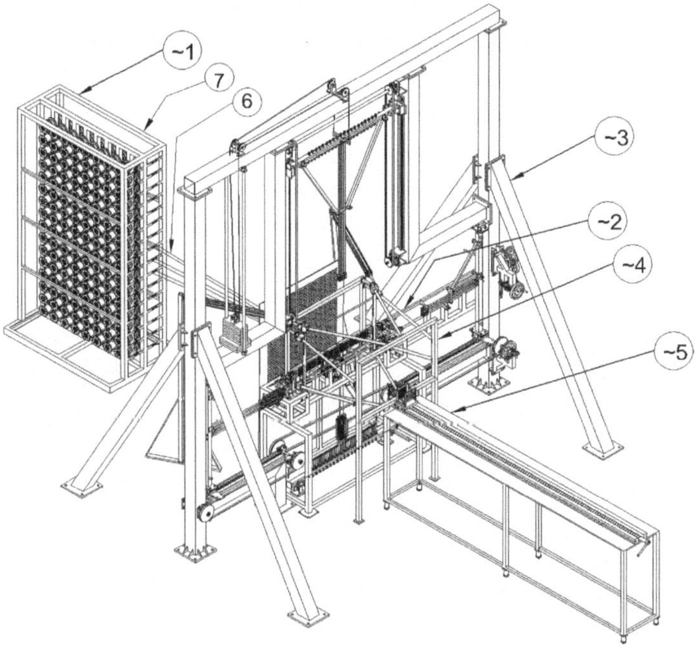

Fig. 13 Isometric view of the 3D weaving machine. Parts list: 1: let-off assembly, 2. Shedding assembly, 3. Picking assembly, 4. Beat-up assembly, 5. Take-up assembly, 6. 'X' Yarn, 7. Creel

4. Design of vacuum-based shuttle transfer mechanism
5. Design of primary and secondary beat-up for compacting inserted yarns
6. Linear take-up for retaining the perform architecture.

Figures 13, 14, 15, 16, 17, 18, 19, 20, 21, 22, 23, 24, 25, 26, 27, 28, 29, 30, 31, 32 and 33 details about the drawings of the machine and its sub-elements. In the drawings the reference numbers indicate corresponding parts in the various figures.

6 Description of the Prototype 3D Weaving Machine (Designed and Developed at CSIR–NAL)

The weaving machine detailed comprise of a let-off assembly (1), a shedding assembly (2), a picking assembly (3), a beat-up assembly (4) and a take-up assembly (5) for producing the woven 3D interlaced perform (173).

The let-off assembly provided for the supply of 'X' yarns (6) is essentially consisting of a creel (7) having let-off of 'X' yarns and a congregate sheet (8) with equally spaced grid holes (9) bonded with ceramic eye-let (10) for arranging the 'X' yarns in a grid-like manner.

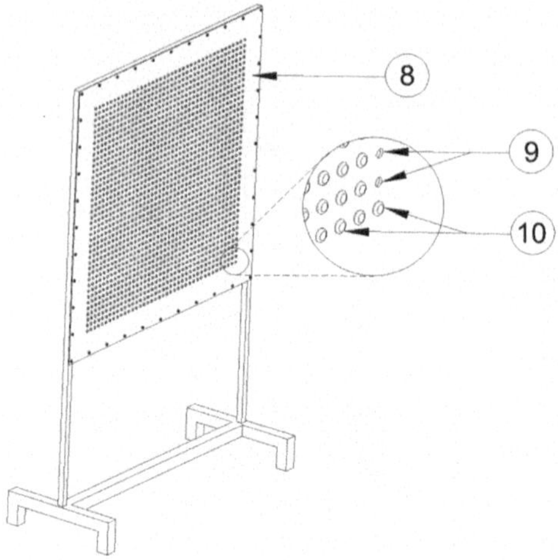

Fig. 14 Isometric view of the congregate sheet. Parts list: 8. Congregate sheet, 9. Equally spaced grid holes, 10. Bonded ceramic eyelets

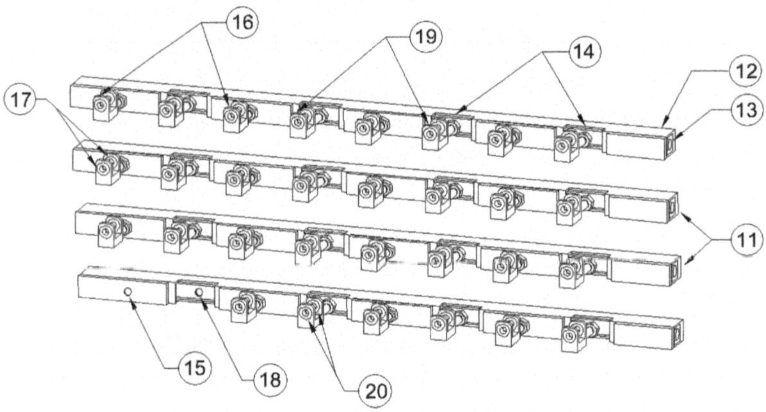

Fig. 15 Outer rectangular tube and inner shaft arrangement in shedding assembly (Home position). Parts list: 11. Series of shedding shafts, 12. Outer Rectangular tube, 13. Inner shaft, 14. Series of equally spaced slits, 15, 18. Equally spaced threadings, 16. Short heald lever, 17, 20. Pair of eye-lets, 19. Long heald lever

The shedding assembly provided for creating horizontal and vertical sheds consists of series of shedding shafts (11) positioned one above the other. The shedding shaft comprises an outer rectangular tube (12) and inner shaft (13). The outer rectangular tube comprises of series of equally spaced slits (14) on the front face along the entire length and equally spaced threadings (15) on the front face in between the said slits. Each threading has a short heald lever (16) mounted to it to guide the 'X' yarn. The short heald lever has a pair of eye-lets (17) spaced apart. The inner shaft has equally

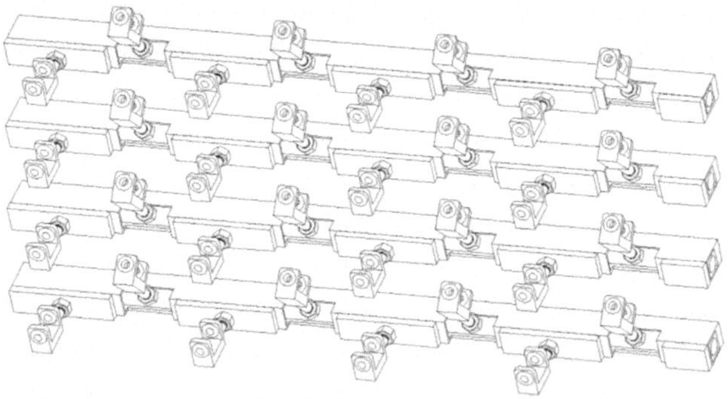

a. Shed open position by clockwise rotation of outer square tube

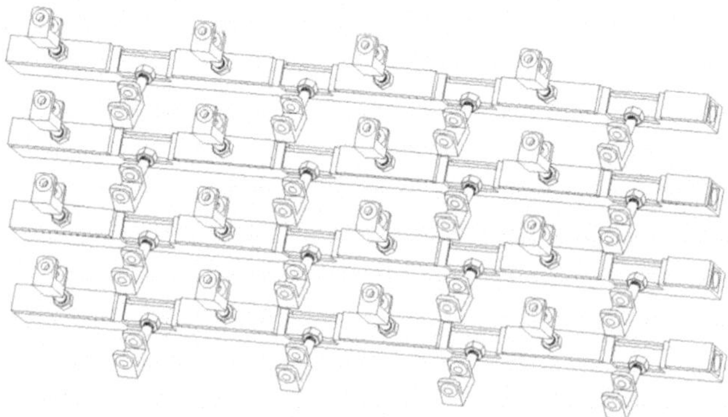

b. Shed open position by anti-clockwise rotation of outer square tube

Fig. 16 Vertical shedding for horizontal weft insertion

spaced threadings (18) on the front face along the entire length centered to the said slits of the outer rectangular tube. The threading houses a long heald lever (19) to guide the 'X' yarn. The long lever has a pair of eyelets (20) spaced apart. The long heald levers project out of the slits of the outer rectangular tube to the same extent as that of the said short heald levers mounted on the outer rectangular tube. The outer rectangular tubes are connected to the main shaft (21) through a link assembly. The link assembly is connected to the said main shaft using an 'H' link (22). The 'H' link connects to the front and back long driver links (23 and 24) through pins (25) at either end. The front long driver link is connected to a driven horizontal link (26) through a cantilever pin (27). The driven horizontal link is connected to an auxiliary horizontal link (28) through a vertical link (29) and pins (30), the other end of the said driven and auxiliary horizontal links are connected to odd outer rectangular tubes. The back long driver link is connected to a driven horizontal link (31) through a cantilever pin (32). The driven horizontal link is connected to an auxiliary horizontal link (33) through a vertical link (34) and pins (35), the other end of the said driven and auxiliary horizontal links are connected to even outer rectangular tubes. The main shaft is mounted on a shedding frame (36) using pillow block (37) and square

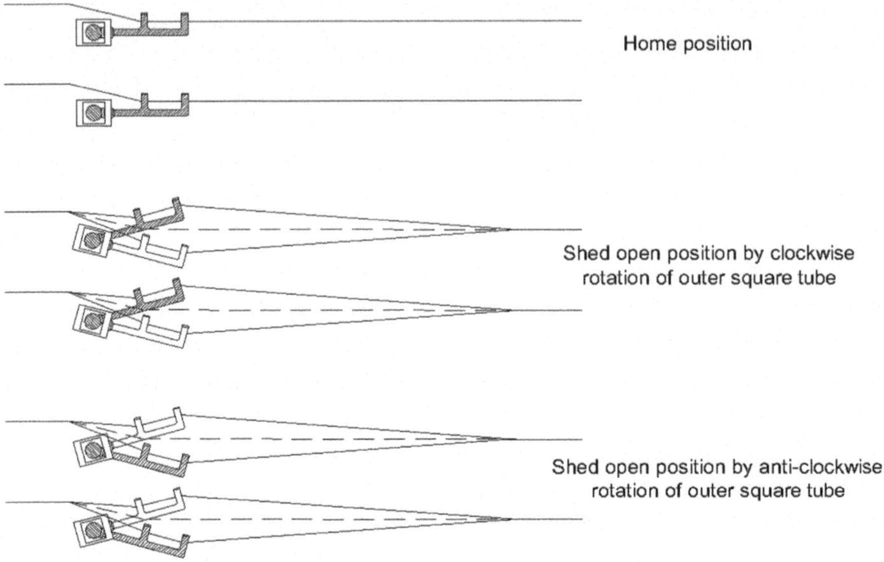

Fig. 17 Side view of vertical shedding

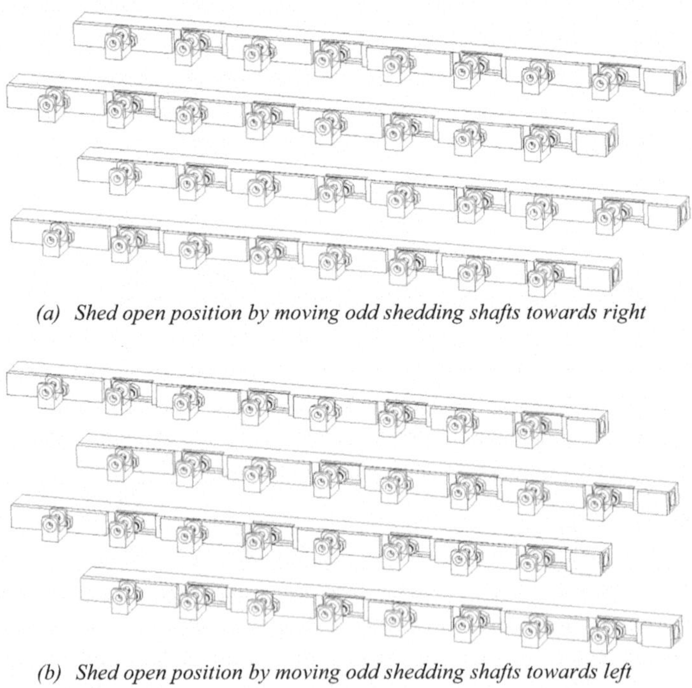

(a) Shed open position by moving odd shedding shafts towards right

(b) Shed open position by moving odd shedding shafts towards left

Fig. 18 Horizontal shedding for vertical weft insertion

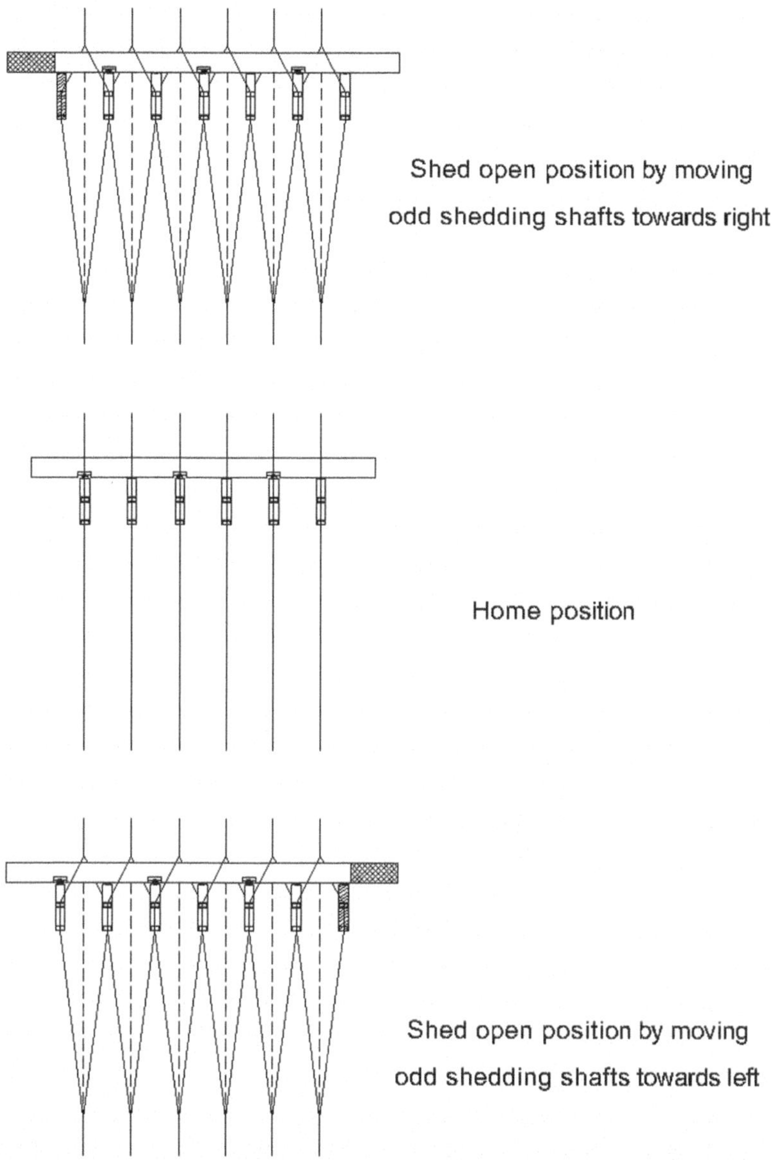

Fig. 19 Top view of horizontal shedding

flange units (38 and 39). The main shaft has a shedding hand lever (40) at one end with stoppers (41) to restrict the movement. The main shaft connects to a reversing shaft (42) at the other end through bevel gears (43, 44 and 45). The bevel gear (44) is mounted on the said shedding frame using a short shaft (46) and a square flange unit (47). The reversing shaft is mounted on the said shedding frame using square

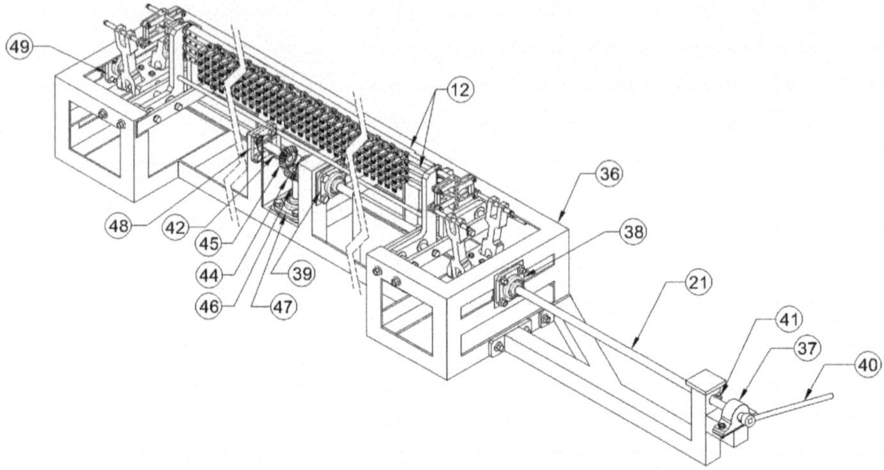

Fig. 20 Drive for vertical shedding. Parts list: 12. Outer Rectangular tube, 21. Main shaft, 36. Shedding frame, 37. Pillow block, 38, 39. Square flange units, 40. Hand lever, 41. Stopper, 42. Reversing shaft, 44, 45. Bevel gears, 46. Short shaft, 47, 48, 49. Square flange units

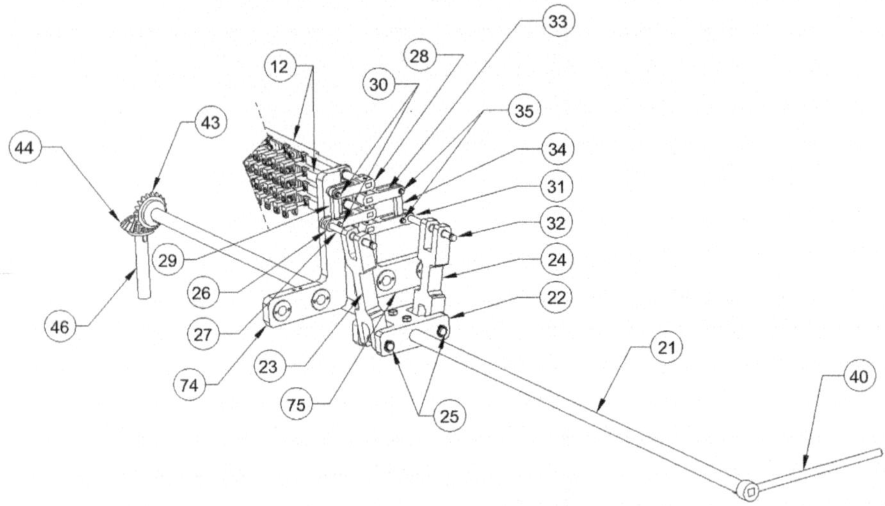

Fig. 21 Right side link arrangement for vertical shedding. Parts list: 12. Outer Rectangular tube, 21. Main shaft, 22. 'H' Link, 23, 24. Front and back long driver links, 25, 30, 35. Pins 26, 31. Driven Horizontal link, 27, 32. Cantilever pin, 28.33 Auxiliary horizontal link, 29, 34. Vertical link, 40. Hand lever, 43, 44. Bevel gears, 46. Short shaft, 74, 75. Backsliding brackets

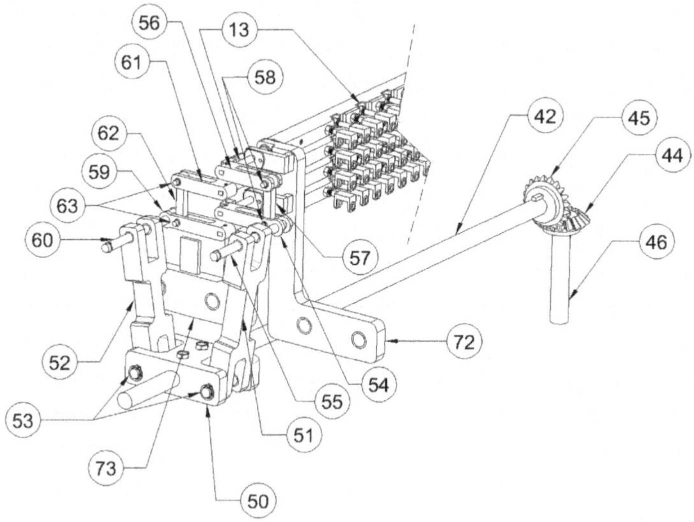

Fig. 22 Left side link arrangement for vertical shedding. Parts list: 13. Inner shaft, 42. Reversing shaft, 44, 45. Bevel gears, 46. Short shaft, 50. H Link, 51, 52. Back long driver links, 53, 58, 63. Pins, 54, 59. Driven horizontal link, 55, 60. Cantilever pin, 56. 61. Auxiliary horizontal link, 57. 62 Vertical link, 72, 73. Front & back sliding brackets

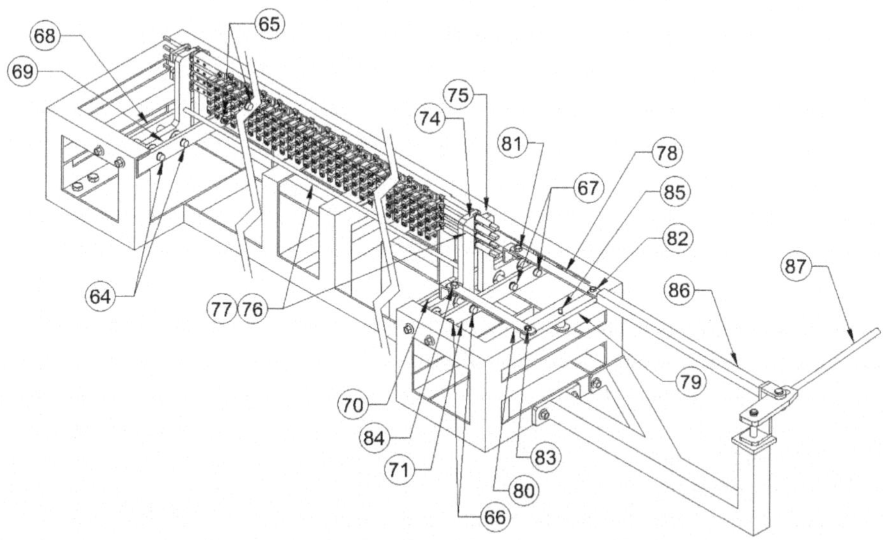

Fig. 23 Drive for horizontal shedding. Parts list: 64, 65, 66, 67, 81, 82, 83, 84, 85 Pins, 68, 69, 70, 71. Flats, 74, 75. Front & back sliding brackets, 76, 77. Connecting rods, 78, 79, 80. Horizontal levers of four-bar mechanism, 86. Long rod, 87. Hand lever

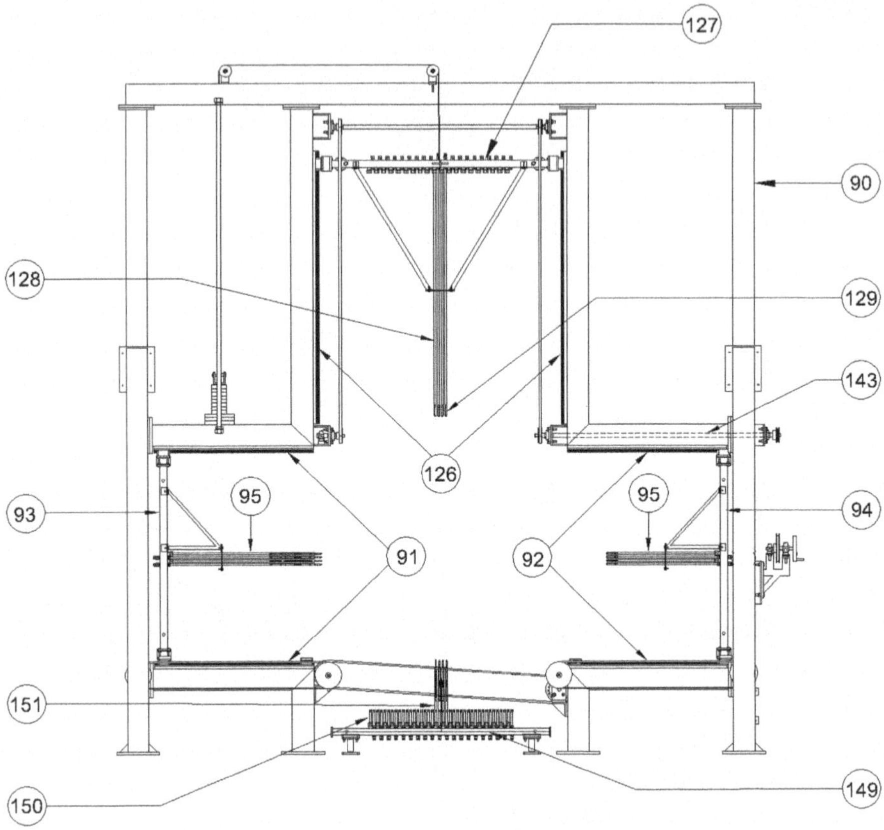

Fig. 24 Front view of picking assembly. Parts list: 90. Picking frame, 91, 92, 126. Linear Motion Rails, 93, 94. Vertical Square tubes, 95. Horizontal picking pipe, 127, 149. Horizontal Square tubes, 128, 150. Vertical picking pipes, 129, 151. Shuttle holder, 143. Long shaft

flange units (48 and 49). The reversing shaft connects to inner shafts through a link assembly. The link assembly is connected to the said reversing shaft using an 'H' link (50). The 'H' link connects to the front and back long driver links (51 and 52) through pins (53) at either end. The front long driver link is connected to a driven horizontal link (54) through a cantilever pin (55). The driven horizontal link is connected to an auxiliary horizontal link (56) through a vertical link (57) and pins(58). The other end of the said driven and auxiliary horizontal links are connected to odd inner shafts. The back long driver link is connected to a driven horizontal link (59) through a cantilever pin (60). The driven horizontal link is connected to an auxiliary horizontal link (61) through a vertical link (62) and pins (63). The other end of the said driven and auxiliary horizontal links are connected to even inner shafts. The shedding frame consists of four pairs of pins(64, 65, 66 and 67) mounted horizontally on flats (68, 69, 70 and 71). The pairs of pins house front and back sliding brackets (72, 73, 74 and 75) on both sides. The front and back sliding brackets (72 & 73) are alternatively

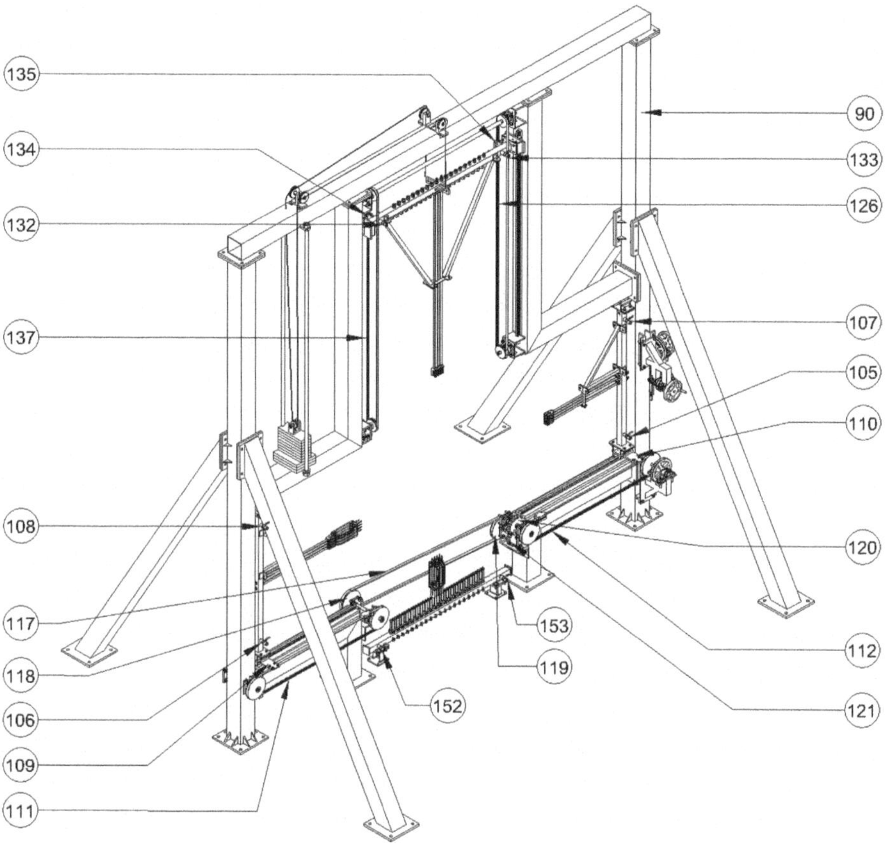

Fig. 25 Isometric view (from left side) of picking assembly. Parts list: 90. Picking frame, 105, 106, 132, 152. Inlet valves, 107, 108, 133, 153. Outlet Valves, 109, 110, 134, 135. Extended plates, 111, 112. Timing pulleys, 117, 137. Timing belts, 118, 119. Pulleys, 120, 121. Gear train, 126. LM Rails

connected to successive inner shafts. The front and back sliding brackets (74 and 75) are alternately connected to successive outer rectangular tubes. The front sliding brackets (72 and 74) are linked together by a connecting rod (76). The back sliding brackets (73 and 75) are linked together by another connecting rod (77). The flats (70 and 71) act as a stopper for sliding brackets. The front and back sliding brackets (74 and 75) are connected to a four-bar mechanism consisting of three horizontal levers (78, 79 and 80) connected by pins (81, 82, 83 and 84) arranged perpendicular to each other in the 'U' shape. The lever (79) is pivoted to said shedding frame using a pin (85). The four-bar mechanism is connected to a long rod (86) mounted on the said shedding frame through the pin (82). The long rod has a hand lever (87) at the other end.

The picking assembly provided for insertion of 'Y' and 'Z' yarns (88 and 89) in the vertical and horizontal sheds consists of a picking frame (90) having horizontal

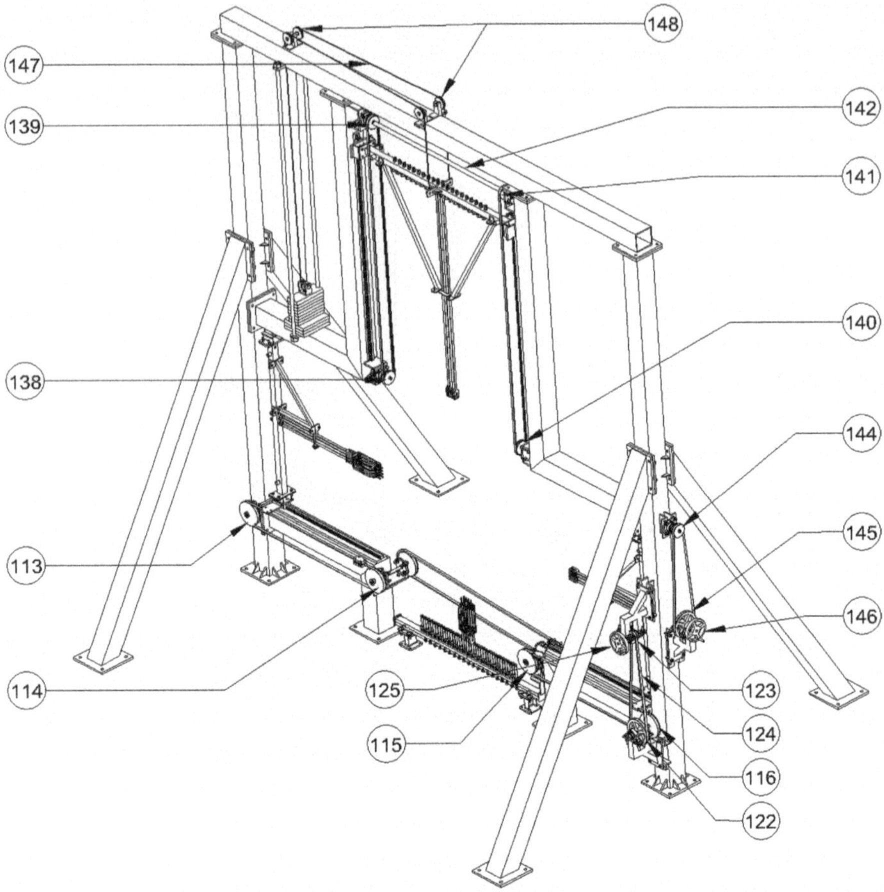

Fig. 26 Isometric view (from right side) of picking assembly. Parts list: 113, 114, 115, 116, 138, 139, 140, 141. Timing Pulleys, 122, 123, 124, 144, 145,. V Belt Pulleys, 125. Picking Handle, 142. Shaft, 146. Shaft with handle, 147. Guide rope, 148. Guide pulleys

and vertical picking systems. The horizontal picking system consists of two pairs of Linear Motion (LM) rails (91 and 92) mounted horizontally on the left and right sides of the said picking frame. The LM rails guide vertical square tubes (93 and 94) consisting of equally spaced horizontal picking pipes welded to it. Each horizontal picking pipe (95) is mounted with a shuttle holder (96). Each shuttle holder consists of two holes (97) facing the shedding assembly to hold the shuttle (98). The shuttle comprises a long rod (99) to hold the pirn (100) and a spring as the tensioning device (101). The pirn has 'Y' yarn wound on it. The shuttle has a slot (102) on the front to guide the yarn during weaving. The shuttle has a pair of pins (103) on either side, which coincide with the said holes of the said shuttle holder. The tensioning device comprises an adjustable nut (104) threaded onto the long rod to vary the tension of the unwinding yarn during weaving. The vertical square tubes are connected to a vacuum

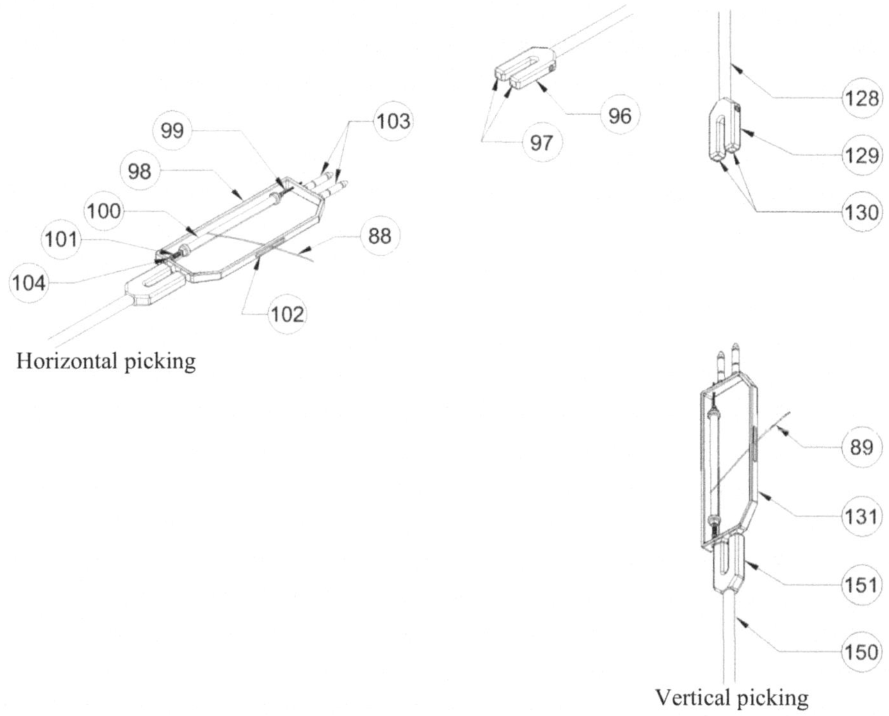

Fig. 27 Shuttle arrangement in horizontal and vertical picking assemblies. Parts list: 88, 89. Y, Z yarn, 96, 129, 151. Shuttle holders, 97, 130. Holes in the shuttle holder, 98, 131. Shuttles, 99. Rod to hold pirn, 100. Pirn, 101. Tensioning device, 102. Slot, 103. Pair of pins, 104. Adjustable nuts, 128. Vertical picking pipe, 150. Small picking pipe

source through inlet valves (105 and 106). The vertical square tubes have outlet valves (107 and 108) as vacuum vents. The vertical square tubes have extended plates (109 and 110) clamped to timing belts (111 and 112). The timing belts are mounted onto the said picking frame by means of timing pulleys (113, 114, 115 and 116). These timing belts on the left and right sides are linked together through a third timing belt (117) with pulleys (118 and 119) mounted on the picking frame. The timing belt ensures synchronized and simultaneous movement of vertical square tubes towards and away from the shedding assembly through a gear train (120 and 121). The timing pulley (116) and a 'V' belt pulley (122) are mounted on the same shaft. The 'V' belt pulley is connected to a smaller 'V' belt pulley (123) by means of another 'V' belt (124). The smaller 'V' belt pulley is mounted on a shaft having a picking handle (125) for rotation. The vertical picking system consists of a pair of LM rails (126) mounted vertically on the top of the picking frame. The LM rails guide a horizontal square tube (127) consisting of equally spaced vertical picking pipes welded to it. Each vertical picking pipe (128) is mounted with another shuttle holder (129) consisting of two holes (130) facing the shedding assembly to hold another shuttle (131) consisting of

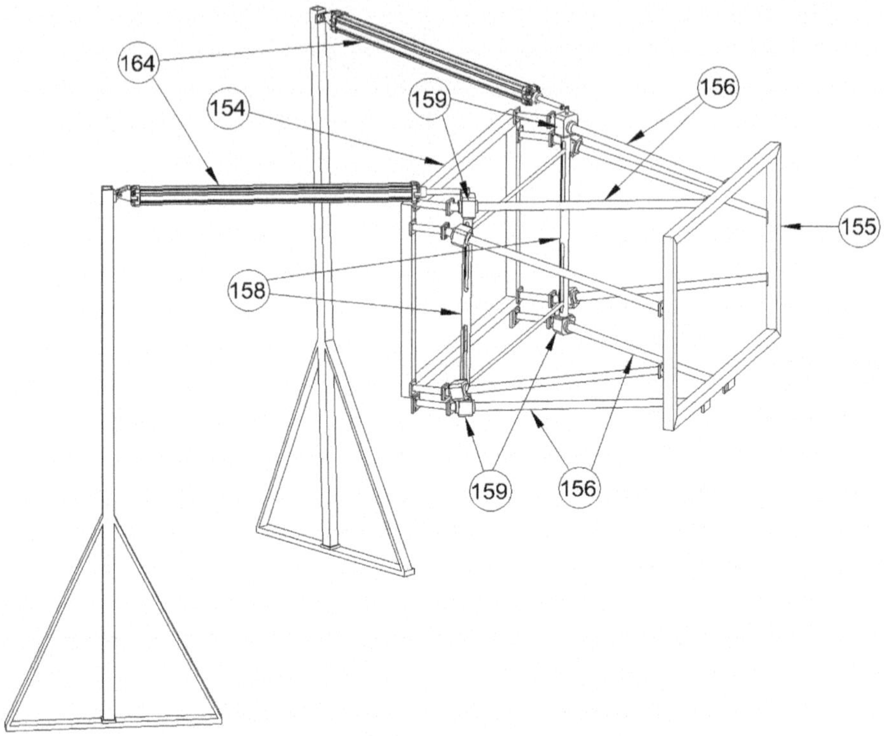

Fig. 28 Isometric view of beat-up assembly. Parts list: 154, 155 Rectangular frames, 156. Guide rods, 158. Vertical beat-up rods, 159. Linear ball bearings, 164. Pneumatic cylinders

Fig. 29 Beat-up assembly with beat-up rods in front position. Parts list: 154, 155. Rectangular frames, 157. Guide rods, 160. Beat-up rods, 161. Linear ball bearings

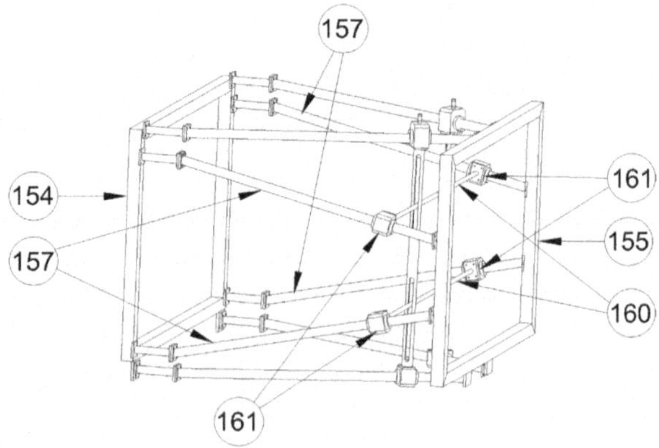

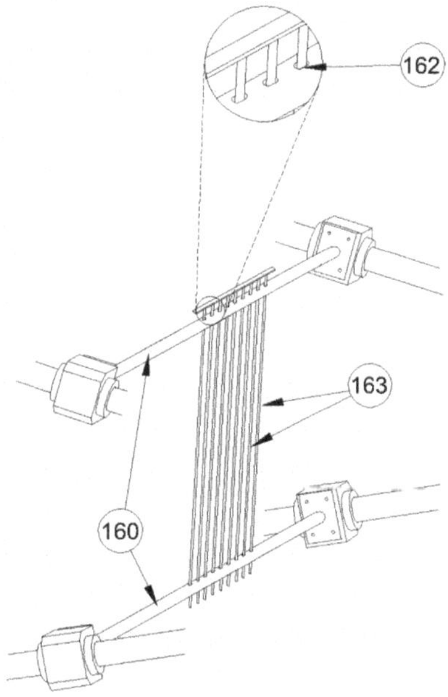

Fig. 30 Sub-elements of the secondary beat-up arrangement. Parts list: 160. Horizontal Beat-up rods, 162. Equally spaced slots, 163. Flat heald wires

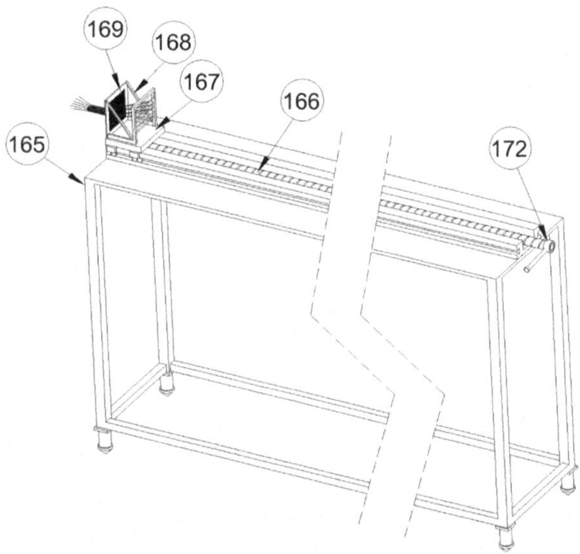

Fig. 31 Isometric view of take-up assembly. Parts list: 165. Take-up frame, 166. Horizontally mounted lead screw, 167. Flat plate, 168. Small frame, 169. Take-up sheet, 172. Handle

'Z' threads. The horizontal square tube is connected to a vacuum reservoir through an inlet valve (132). The horizontal square tube has an outlet valve (133) for the vacuum vent and has extended plates (134 and 135) clamped to timing belts (126 and137) on both sides. Each timing belt is mounted onto the said picking frame by means of timing pulleys (138, 139, 140 and 141). The timing pulleys (139 and 141)

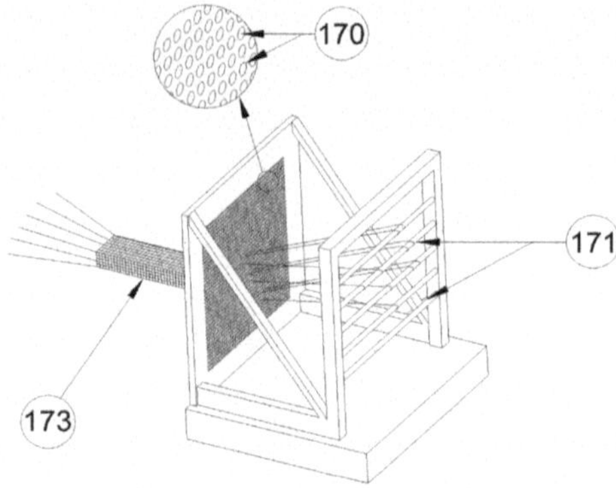

Fig. 32 Isometric view of take-up sheet arrangement. Parts list: 170. Closely spaced holes, 171. Rods, 173. Interlaced preform

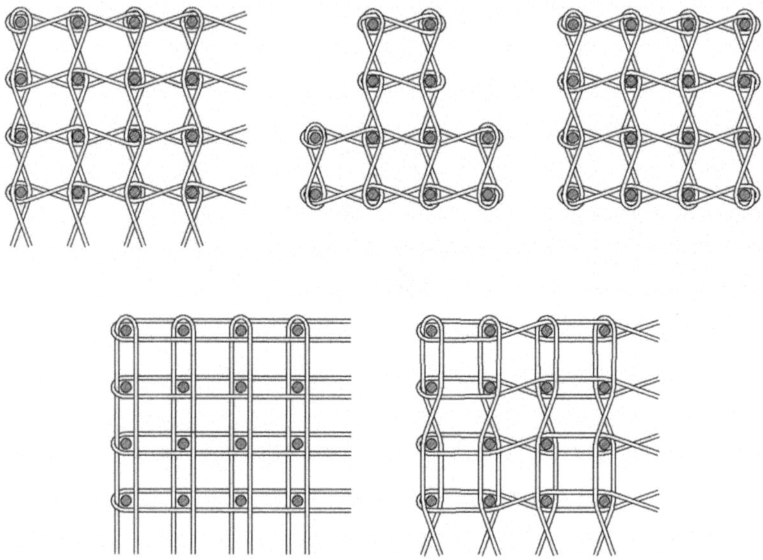

Fig. 33 Possible architectures that can be developed using the 3D weaving machine

are linked together by means of a shaft (142). The timing pulley (140) is mounted on a long shaft (143). The long shaft passing through the said picking frame on the other end consists of a small 'V' belt pulley (144). The small 'V' belt pulley connects to a large 'V' belt pulley (145) mounted on a shaft having a handle (146). The horizontal square tube is connected to an adjustable counterweight system through a wire rope (147) and guide pulleys (148). The picking frame has another horizontal square tube positioned at the bottom (149). The horizontal square tube has equally spaced small picking pipes (150) welded to it. Each picking pipe is mounted with a shuttle holder

(151) consisting of two holes facing the shedding assembly to hold the shuttle. The horizontal square tube is connected to the said vacuum reservoir through a pair of inlet and outlet valves (152 and 153).

The beat-up assembly provided for the compaction of weft yarns consists of two rectangular frames (154 and 155). The rectangular frame (154) is positioned in front of the said shedding assembly and behind the said horizontal and vertical picking pipes. Another rectangular frame (155) is positioned near the take-up assembly. The rectangular frame (154) has eight guide rods (156 and 157) welded to it at different points along its perimeter. The guide rods are welded to the said rectangular frame (155) as per the required converging angle of the 'X' yarns. The guide rods (156) are linked together using vertical beat-up rods (158). The vertical beat-up rods are mounted on the guide using pillow blocks with linear ball bearings (159). The horizontal beat-up rods have two long slots through which horizontal beat-up rods (160) pass. The horizontal beat-up rods are mounted onto guide rods (157) using pillow blocks with linear ball bearings (161). The horizontal beat-up rods have equally spaced slots (162) to hold flat heald wires (163). The vertical beat-up rod (top one) is linked to two pneumatic cylinders (164) at either end. The pneumatic cylinders are placed at the same line path of the converging guide rods near the said congregate sheet.

The take-up assembly provided for the retention of woven performs consists of a take-up frame (165). The take-up frame has a horizontally mounted lead screw (166). The lead screw has a flat plate (167) mounted on it. The flat plate is attached to a small frame (168) consisting of a take-up sheet (169) mounted perpendicular to it and parallel to said congregate sheet. The take-up sheet has closely-spaced holes (170) arranged in a grid-like manner to receive the said 'X' yarns from the said shedding assembly. Each hole in the said take-up sheet corresponds to one 'X' yarn. The 'X' yarns passing through the said hole are tied to rods (171) welded horizontally on the said small frame. The lead screw has a handle (172) at one end to move the small frame.

The warp yarn is fed from the creel or beams with the indirect or direct tensioning method and beat-up assembly synchronized with take-up and let-off assemblies for achieving better preform consistency.

The 3D weaving machine driven by either mechanical or electronic means is capable of generating any specific profile like T, C, I, H, π, + , X etc. within the given grid dimensions using glass, carbon, Kevlar, silk, cotton and the like or combination thereof.

The fibre content of 3D interlaced structure is increased with the insertion of non-interlaced stuffer yarns in the axial direction.

7 Working of the Machine

The machine produces 3D fabrics of block configuration and/or specific types of profiles. This is achieved by adopting the operation of four weaving cycles as disclosed below. Each weaving cycle comprises of sequential operation of the sub-elements of the weaving machine. The sequential operation comprises of shedding in the desired plane, followed by corresponding weft insertion, further followed by primary and secondary beat-up and finally take-up in conjunction with let-off activities.

1. Weaving cycle 1 for horizontal weft insertion from right to left
2. Weaving cycle 2 for vertical weft insertion from bottom to top
3. Weaving cycle 3 for horizontal weft insertion from left to right
4. Weaving cycle 4 for vertical weft insertion from top to bottom.

To complete one unit cell of a 3D block structure the order of the weaving cycle would be 1, 2, 3 and 4 or 1, 3, 2 and 4. To complete one unit cell of a 3D profile such as 'T' or 'I' the order of the weaving cycle would be 1, 3, 2 and 4. Repetitive weaving of the unit cell would give the required length of 3D preform. Now the working is described in terms of the above four weaving cycles.

Prior to the start of any weaving cycle, the 'X' threads from individual bobbins mounted on the creel are drawn through the congregate sheet and through the respective heald levers of the shedding assemblies. The number of threads and their arrangement is decided by the type of 3D fabric that needs to be woven. The threads are tied to the frame of the take-up through the respective holes of the take-up sheet. The heald levers are in a level position, where all the heald levers mounted on each shedding shaft assembly (inner shaft and outer tube) are aligned in one plane. The heald levers of successive shaft assemblies are kept parallel to each other in their respective planes. The operating handles for the horizontal and vertical shedding are in their respective home positions.

7.1 Weaving Cycle 1: Horizontal Weft Insertion from Right to Left

Vertical shedding for horizontal weft insertion is obtained by partially rotating the shedding handle in the clockwise direction from its level position. This movement rotates the long shaft partially. The partial rotation of the shaft swivels the 'H' link pushing the front long link up and back long link down through the respective pins. The front long link gives an upward motion to the primary driven and auxiliary driven links through the pin. This upward motion of links gives a partial rotation to the alternate outer tubes in the clockwise direction. Simultaneously, the back long link provides partial clockwise rotation to the other set of alternate outer tubes through the primary driven, auxiliary driven links and pin. The partial rotation of the outer tube displaces the projected end of all the heald levers mounted on them to half the shed depth in the upward direction. The warp thread passing through these heald levers deviates from their straight path by half shed depth. A stopper is provided to restrict the movement of heald levers beyond the predefined shed depth.

The partial rotation of the long shaft in the clockwise direction simultaneously results in an equal amount of anticlockwise rotation of the short reversing shaft by the bevel gear assembly. This anticlockwise rotation of the short reversing shaft swivels the ‘H’ link in the opposite direction pushing the front long link down and back long link up through the respective pins. The front long link in turn pulls the primary driven and auxiliary driven links, as a result of which the alternate inner shafts are partially rotated in the anticlockwise direction. Simultaneously, the back long link provides partial anticlockwise rotation to the other set of alternate inner shafts through the pin and primary driven and auxiliary driven links. The partial rotation of the inner shaft displaces the projected end of all the heald levers mounted on them to half the shed depth in the downward direction. The warp passing through these heald levers deviates from their straight path by half shed depth. This simultaneous creation of the upward and downward half sheds by alternative warp threads provides the required shed depth for horizontal weft insertion from right to left. The length of the picking shaft is about 1000 mm. The max depth of the shaft cannot exceed 17 mm and the deflection with the shuttle should not exceed 1.00 mm. This is almost impossible to achieve due to the overhang length of 1000 mm. To circumvent this problem, the tubes were split into two halves of each 500 mm along the lengthwise and the shuttle transfer takes place at the middle of the weaving cycle. This concept helped in limiting the deflection within 1 mm.

The weight and stiffness of the shuttle also plays an important role to reduce the deflection of the picking shaft. To achieve this, the shuttle is made out of carbon composites.

Prior to horizontal weft insertion, shuttles comprising of ‘Y’ yarns wound on pirns are held by means of vacuum on the right-hand side (handle side) of the machine. Rotating the picking handle in an anti-clockwise direction moves both the vertical square tubes of the horizontal picking system towards the warp layers. Further movement of the handle results in the shuttle entering the respective warp sheds. At the centre of the picking plane the shuttles are transferred from the right set of picking pipes to the left set of picking pipes guided by the pins on the respective shuttles. Operation of respective vacuum valves results in release and suction of vacuum on the right and left picking pipes, respectively. This in turn results in the release of the shuttles from the picking pipes on the right and are now held by the suction force of vacuum by the left picking pipe.

The handle is rotated in the clockwise direction which results in the shuttle moving away from the handle towards the left of the machine, out of the respective sheds. The transfer of shuttle from right to the left results in laying-out of the ‘Y’ yarns through the respective slots of the shuttle into the sheds. This laying-out of weft to the required length is regulated by the tensioning arrangement in the shuttle.

The position of the vertical beat-up rods is now behind the just inserted weft. The primary vertical beat-up rods actuated by pneumatic cylinders push the just inserted horizontal wefts towards the fell of the 3D fabric. Secondary vertical beat-up rods compact these wefts to the fell.

The take-up plate now moves by one pick spacing. The movement of the take-up plate draws the 'X' threads by an equal amount from the creel regulated by the tensioning arrangement on the individual warp bobbin.

After beat-up of the horizontal wefts, the handle is moved back to the home position which results in levelling of the heald levers in the horizontal plane. Now the beat-up rods are retracted back to their stationary position near the shedding assemblies.

7.2 *Weaving Cycle 2: Vertical Weft Insertion from Bottom to Top*

Pulling the shedding handle away from the shedding assembly pulls the long arm. The long arm actuates the four-bar mechanism in such a manner that the front sliding bracket is pushed to the left of the machine and backsliding bracket is pulled to the right of the machine. The front sliding bracket moves the connected alternate shedding shaft assemblies to the left of the machine by an equal amount in the vertical plane. The projected end of the heald levers on these shaft assemblies move by half the shed depth to the left from the level position of the healds. The warp threads threaded through the heald levers in turn are displaced by half the shed depth in the horizontal plane to the left. The backsliding bracket moves the alternate set of shedding shaft assemblies to the right of the machine by half the shed depth in the horizontal plane. The heald levers mounted on these shafts also move by half a shed depth to the right which in turn displaces the warp threads threaded through the heald levers by half shed depth to the right. This simultaneous movement of successive shedding shaft assemblies and the warp threads in the opposite direction creates the required shed depth in the horizontal plane for vertical weft insertion.

Prior to vertical weft insertion shuttles comprising of 'Z' yarns wound on pirns are held by means of vacuum on the bottom square tubes of the vertical picking system. Rotating picking handle in a clockwise direction moves the top horizontal square tubes of the vertical picking system towards the warp layers. Further movement of the handle results in the picking pipes entering the respective warp sheds until the shuttle holders project out of the sheds at the bottom and are guided onto the pins of the respective shuttles. Operation of respective vacuum valves results in release and suction of vacuum on the bottom and top picking pipes, respectively. This intern results in the release of the shuttles from the picking pipes on the bottom, and they are now held by the suction force of vacuum by the top picking pipes. The handle is rotated in the anti-clockwise direction which results in the shuttle moving towards the top through the shed layers and out of the respective sheds. The transfer of shuttle from bottom to top results in laying-out of the 'Z' yarns through the respective slots of the shuttle into the sheds. This laying-out of weft to the required length is regulated by the tensioning arrangement in the shuttle.

The position of the horizontal beat-up rods is now behind the just inserted weft. The primary horizontal beat-up rods actuated by pneumatic cylinders push the just inserted vertical wefts towards the fell of the 3D fabric. Secondary horizontal beat-up rods compact these wefts to the fell.

The take-up plate now moves by one pick spacing. The movement of the take-up plate draws the 'X' threads by an equal amount from the creel regulated by the tensioning arrangement on the individual warp bobbin.

After beat-up of the vertical wefts, the handle is moved back to the home position which results in levelling of the heald levers in the vertical plane. Now the beat-up rods are retracted back to their stationary position near the shedding assemblies.

7.3 Weaving Cycle 3: Horizontal Weft Insertion from Left to Right

Weaving cycle 3 is similar to weaving cycle 1 with the following changes. The shedding handle is rotated in an anti-clockwise direction from its level position resulting in partial rotation of the outer tubes in the anti-clockwise direction. This displaces the projected end of all the heald levers mounted on them to half the shed depth in the downward direction. This action of the heald lever displaces the warp yarns in the downward direction from its straight path by half shed depth. The inner shafts are partially rotated in the clockwise direction. This displaces the projected end of all the heald levers mounted on them to half the shed depth in the upward direction. This action of the heald lever displaces the warp yarns up from its straight path by half shed depth.

Horizontal weft insertion is carried out from left-hand side to right-hand side by relevant operations of the picking handle and vacuum valves.

7.4 Weaving Cycle 4: Vertical Weft Insertion from Top to Bottom

Weaving cycle 4 is similar to weaving cycle 2 with the following changes. The shedding handle is pulled towards the shedding assembly. The front sliding bracket is pushed to the right of the machine, and the backsliding bracket is pulled to the left of the machine. This results in reversal of the position of the heald levers as described in weaving cycle 2 from left to right or right to left providing required shed depth.

Weft insertion is carried out as described in weaving cycle 2 from top to bottom with relevant operations of the picking handle and vacuum valves.

Figure 34, shows the photograph of the Dual plane shedding based 3D weaving machine. This was a prototype developed to show the functioning of all the sub-elements and their synchronised working.

Fig. 34 Dual plane shedding based 3D Weaving Machine (prototype)

The advantages of the above approach of 3D weaving are as follows.

1. Integrated 3D structure with yarn interlacements in X, Y and Z directions can be obtained.
2. Possibility exists for custom designing of the preform based on the end-use requirement.
3. Dual Plane shedding for 3D weaving is achieved by the use of counter-rotating and lateral traverse of concentric devices. Basically, a single device assembly achieves all the shedding requirements.
4. Depending on the type of profile being woven, the heald levers can be mounted/demounted as convenient.
5. Structural integrity at the intersection is ensured along with minimum damage to the warp yarns as it does not undergo the twisting and untwisting cycle.
6. Spacing between the warp yarns would be the optimum for shed formation, and therefore the geometrical constraints would be minimized.
7. The stress and strain on warp yarns are uniform and balanced as alternative warp yarns are displaced equally in opposite directions during shed formation.
8. There is also the possibility of weaving continuous length non-interlaced 3D (noobed) structure with the same set-up, by group controlling the concentric devices.
9. Execution of simultaneous weft insertion in multiple layers of each plane.
10. Use of shuttle with tensioning arrangement provides neat selvedges for all the layers.
11. Dual beat-up system ensures optimum compaction.
12. Linear take-up system ensures unaltered yarn paths and preform architecture, especially for profiles.

13. Reduced damage to yarns while weaving, since the yarn path is guided by stainless steel/ceramic surfaces.

8 Photographs of Machine/Sub-Elements and Dual Plane Shedding Based Woven Preforms

Figures 34, 35, 36, 37, 38, 39, 40, 41, 42, 43 and 44 or this section details the photographs of the prototype machine, the relevant AutoCAD models of the sub-elements and photographs of the woven preforms. All the photographs are self-explanatory.

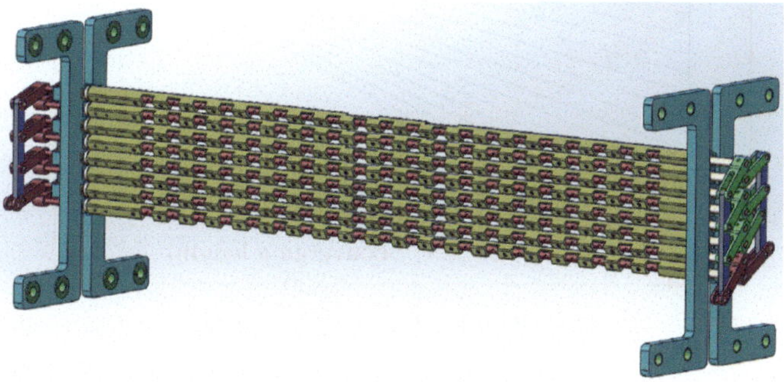

Fig. 35 8 Shaft AutoCAD model of shedding shaft assembly (Top)

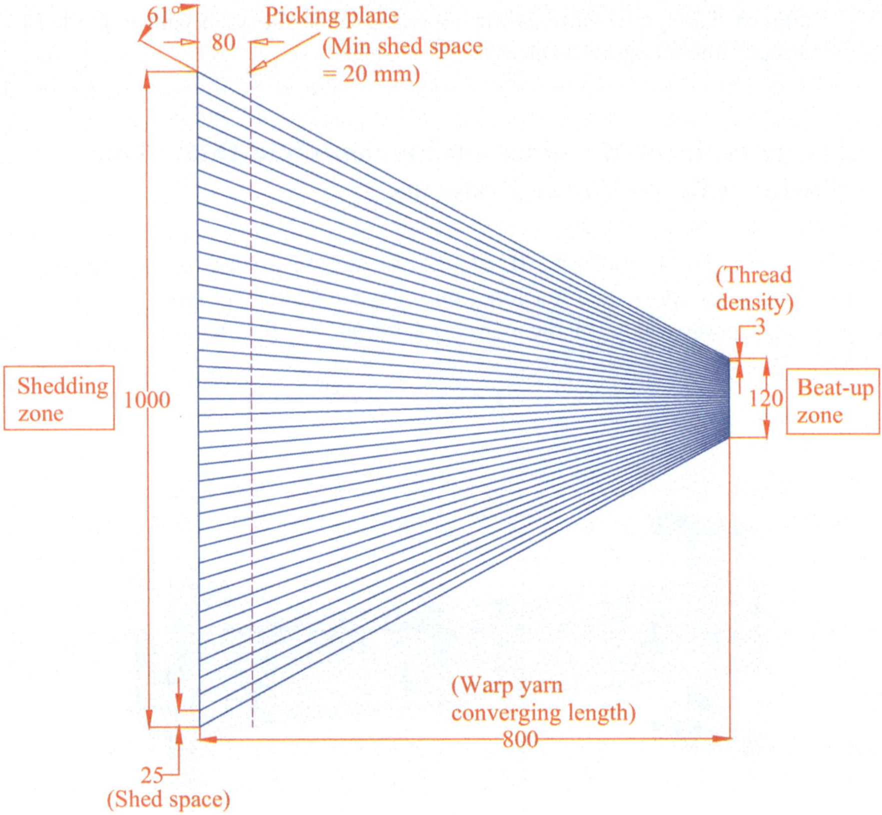

Fig. 36 The Shed spacing diagram used for building the machine

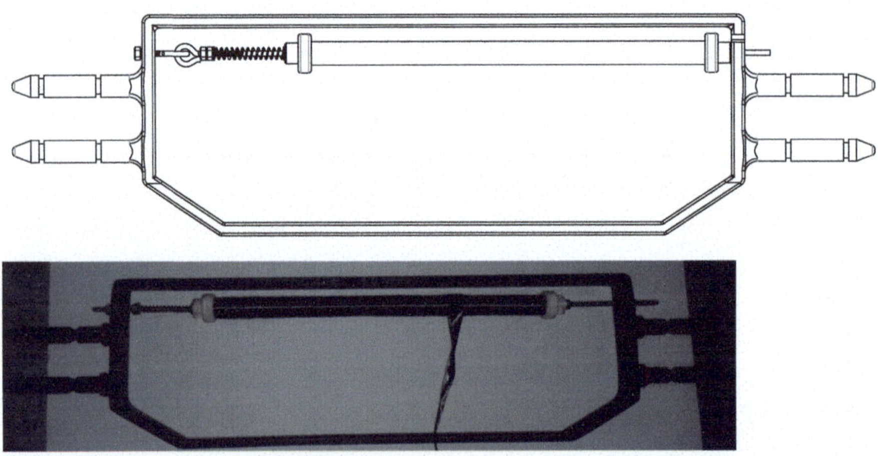

Fig. 37 CFRP shuttle (Top) and the fabricated shuttle with pirn assembly (Bottom)

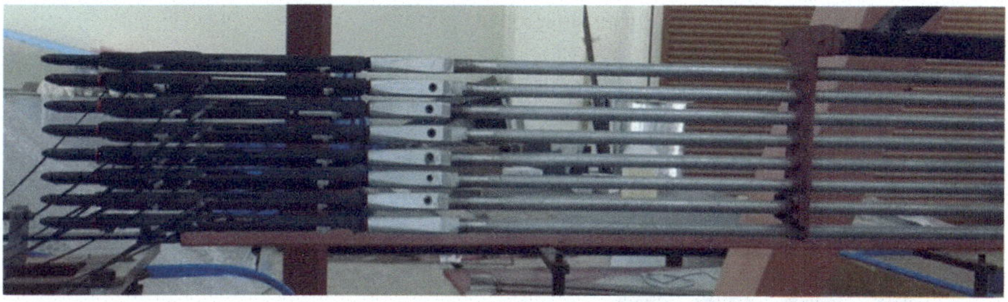

Fig. 38 Vacuum picking devices with CFRP shuttles

Fig. 39 Manufactured primary beat-up assembly

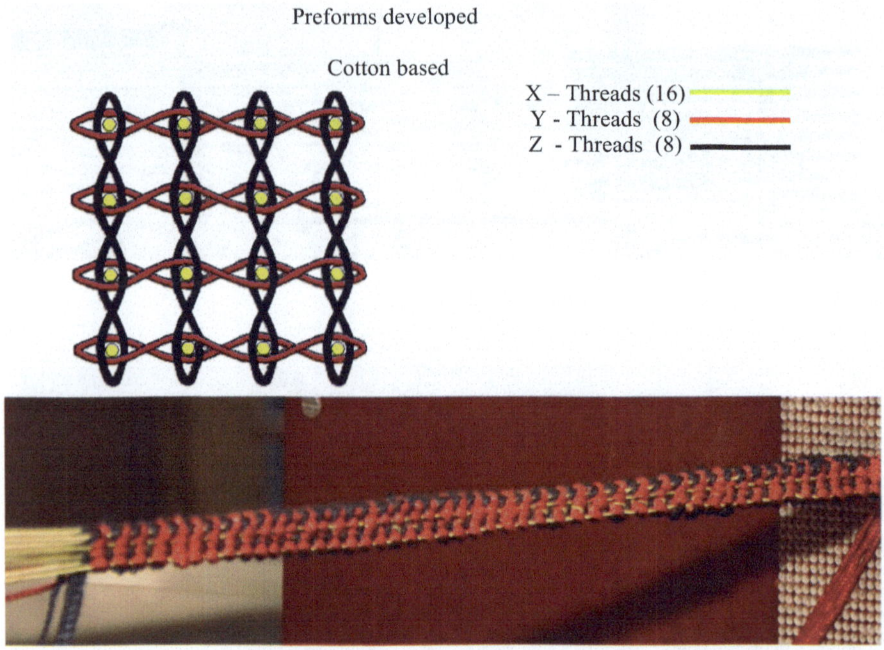

Fig. 40 Weave architecture for cotton sample (Top) and the woven sample (Bottom)

Fig. 41 Weave architecture of carbon-based sample (Top) and the woven sample (Bottom)

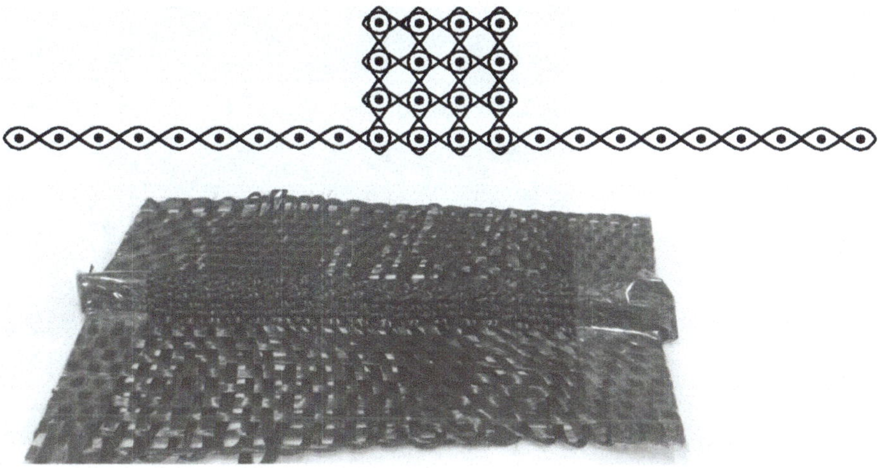

Fig. 42 Weave architecture of T shaped preform (Top) and the woven sample (Bottom)

Fig. 43 Solid 3D woven carbon preform with 24 × 24 mm cross-section. Weave architecture (Top), full-length preform (Middle) and side view of the preform (Bottom)

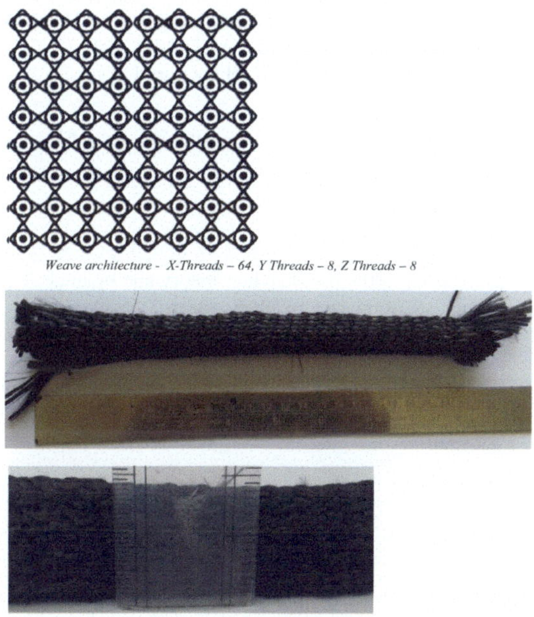

Fig. 44 Interlaced 3D weave pattern (Surface)

Acknowledgements The following students and their institutes are acknowledged for their contributions vide their project works. 1. Nithya Aparna B, ME, Design and development of multiple head carbon roving winding machine. UVCE, Bangalore University, 2012. 2. Revanth Y, Ruthresh Kumar R, Shashikanth Mohan and Shivaprasad B V. BE. Mech Engg. The beat-up mechanism for a 3D weaving Machine. MSRIT, VTU 2015-2016. 3. Giridhar Kalkeri, M E in Machine Design. Design of 3D weaving machine. Bangalore University.2014-2015.

References

1. Khokar N (2001) 3D-Weaving: theory and practice. J Text Inst 92(2):193–207
2. Sugun BS, Ramaswamy S, Sandeep DN, Dayananda GN (2014) An overview on development of three dimensional reinforcements for use in composites at VSSC, Thiruvanathapuram. CSIR-NAL. ISAMPE- INCCOM-13
3. Fredrik S, Stefan H (2009) Assessment of the mechanical properties of a new 3D woven fibre composite material. Compos Sci Technol 1686–1692

Chapter 5
Design Aspects of Dual Plane Shedding-Based 3D Weaving Machine

J. Ramaswamy Setty, Shashikant Choudhary, and B. S. Sugun

1 Introduction

The previous chapter, i.e., Chap. 4 discussed exhaustively the fundamental concepts, machine overview, functioning of the sub-elements and weaving of samples of dual plane shedding-based 3D weaving. However, it does not discuss the design details of the sub-elements which falls under core mechanical engineering discipline. This has been addressed in this chapter. The approach for the design of the sub-elements has been in the adaptation of the well established principles of mechanical engineering to build the prototype, and the discussion of this chapter will be of value addition to the mechanical engineer who wishes to understand the intricacies involved in textile loom design. This chapter does not reiterate on the concepts or working of the machine, since it has been previously covered and the reader is required to refer Chap. 4 for any clarifications regarding the particular sub-element.

2 Configuration of Machine

2.1 Arriving at Machine Dimensions

The first challenge has been to design the framework for the 3D weaving machine. The input used for deciding this frame work was based on the envisaged end-product, viz., preform cross section, viz., 120 × 120 mm. Although the length was not a

J. Ramaswamy Setty (✉) · S. Choudhary · B. S. Sugun
Centre for Societal Missions and Special Technologies (CSMST), CSIR-National Aerospace Laboratories, Bangalore, India
e-mail: ramaswamy@nal.res.in

S. Choudhary
e-mail: Shashi_nal@nal.res.in

B. S. Sugun (ed.), *Practical Approach to 3D Weaving*, Materials Horizons: From Nature to Nanomaterials, https://doi.org/10.1007/978-3-030-80424-4_5

limiting factor, however, it was restricted to 3000 mm. The spacing of the threads at the preform zone was 3 mm, thus needing 40 × 40 grid based warp threads, i.e., 1600 threads. Now diverging from the preform zone to the shedding zone required expansion of the spacing to 30 mm to include the shedding and picking elements. This works out to 40 × 30 = 1200 mm in the horizontal and vertical planes. Using this basic information, detailed design of the machine was arrived at. In addition to design of the sub-elements, the following additional factors were considered prior to arriving at the machine configuration:

- Multiple shedding and picking elements to be designed in a limited space.
- Maintaining uniform tension of the feeder yarn throughout the machine path.
- Maintaining stringent geometric tolerances like Perpendicularity, concentricity, etc., on the machine elements. Any deviation would interfere with adjacent members.
- The Guide and drive for shedding assembly was required to have both Translatory and rotary movements.
- Driving all the shedding elements through series of links from one prime mover.
- Required capability of linkages in shedding to attain reverse motion.
- Bending of unsupported length (1.16 m) of shedding outer tube and inner shaft having minimum clearance b/w each other (1.5 mm).
- The Design of picking device that is capable of transferring the shuttle in a meagre gap of 18 mm.
- Ensuring perfect alignment between top and bottom picking elements over the unsupported length (1.26 m) of vertical picking shafts.
- The Arrangement of picking shafts with allowance of alignment according to the warp angle.
- The unique Design of beat-up mechanism of converging tows in both the planes (vertical as well as horizontal).

Based on the above factors, the machine dimensions worked out to be 10200 × 4400 × 4700 (L × W × H) mm approximately (Fig. 1).

3 Detailed Design Aspects

3.1 Yarn Tension Control

While the textile industry predominantly uses direct tension control approach based on spring tensioner (Fig. 2), the approach adopted for indirect tension control is shown in Fig. 3. Here, the yarn does not come in contact with any tensioner elements. This approach is required for Technical fibres such as carbon to prevent damage and filamentation due to abrasion. The tension here can be controlled by varying the weights.

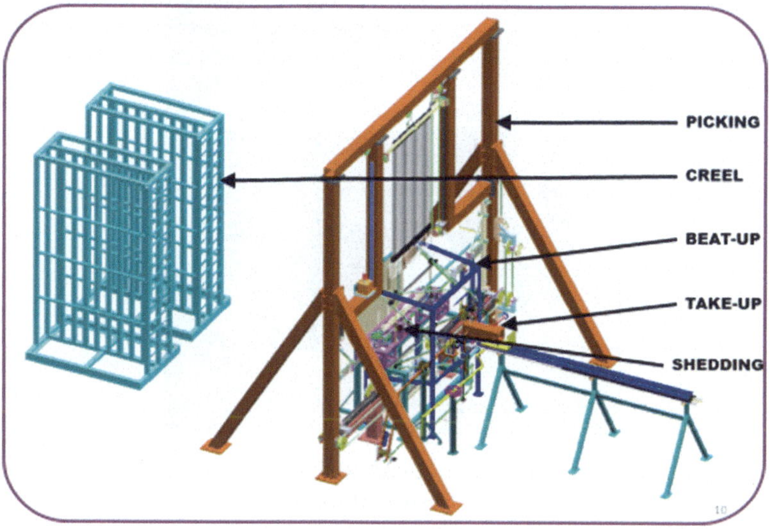

Fig. 1 CAD model of the 3D weaving machine

Fig. 2 Direct tension control approach (Conventional Spring Tensioner)

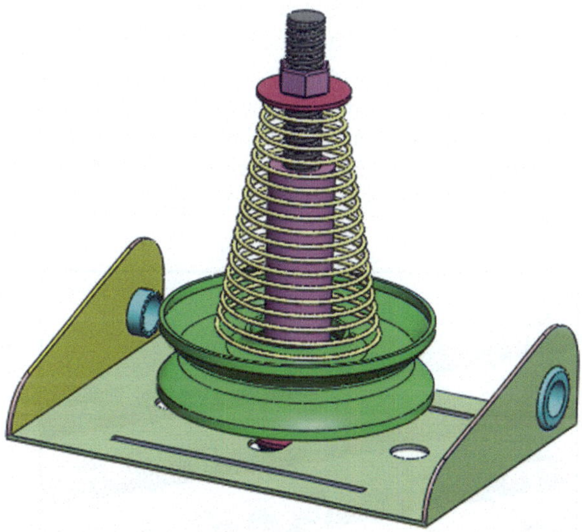

3.2 Shedding System

The shedding shaft consists of two counter rotating shafts with specially designed heald levers to hold the warp mounted alternatively on the solid inner and hollow outer shafts.

Fig. 3 Indirect tension control (Weight based)

To achieve this, initially circular cross section design was adopted for both hollow and solid shaft as shown in the Fig. 4. But during actual fabrication, several difficulties like precise positioning and welding of nuts to hold heald levers on the pipes, circularity of pipes, precision drilling of holes on the pipes in the perpendicular planes, drilling and cutting of threads on the inner shaft in line with the outer pipe etc., were encountered. To circumvent these issues, a square outer pipe concept is adopted as shown in the Fig. 5. In this design, welding of nuts which is the main

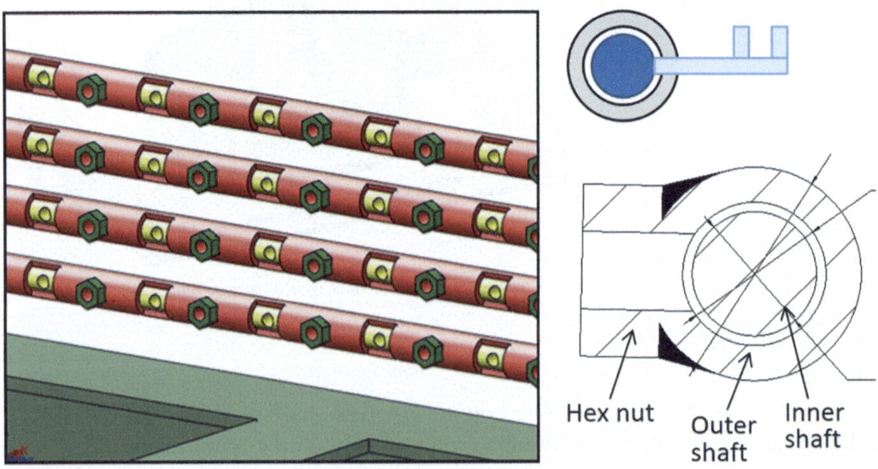

Fig. 4 Circular design of the shedding shafts

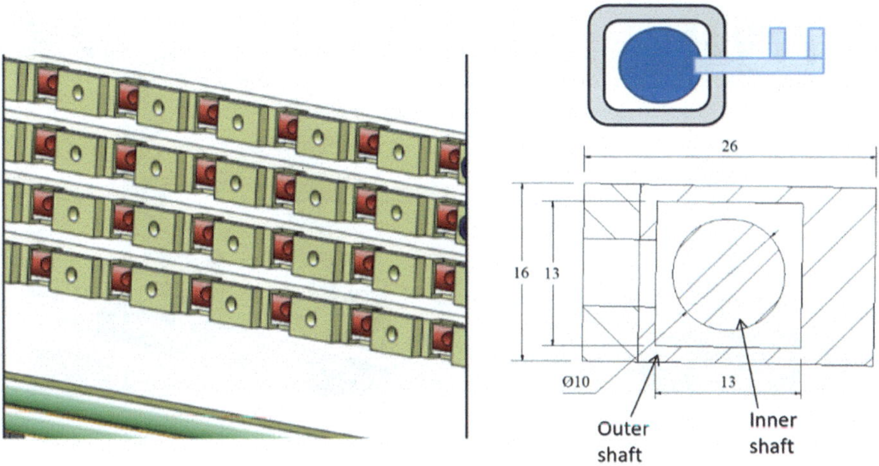

Fig. 5 Square design of the outer shaft

cause for inaccuracy has been avoided. The square pipe has improved stiffness and helps in achieving proper alignment.

Figure 6 shows one full length heald shaft (about 1150 mm). It was observed that due to the self weight a deflection of about 10–12 mm was observed on assembly, which would foul with bottom shaft. To overcome this problem, intermediate supports were provided. But due to the minimal space available between shafts, supports were provided on alternate shafts as shown in Fig. 7.

Although the prototype was conceptualized for 40 × 40 shedding elements, initially 8 × 40 shedding elements were built. The shedding assembly consisted of a shaft inserted in a square tube. In each of the shaft and the square tube, 20 nos of equally spaced tapped holes were drilled, thus making totally 40 horizontal holes in a row when assembled. The arrangement of these holes is such that, the holes on the

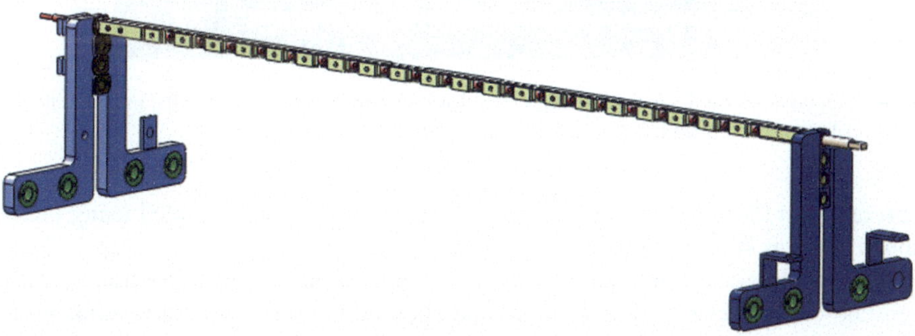

Fig. 6 CAD model of the assembled heald shaft

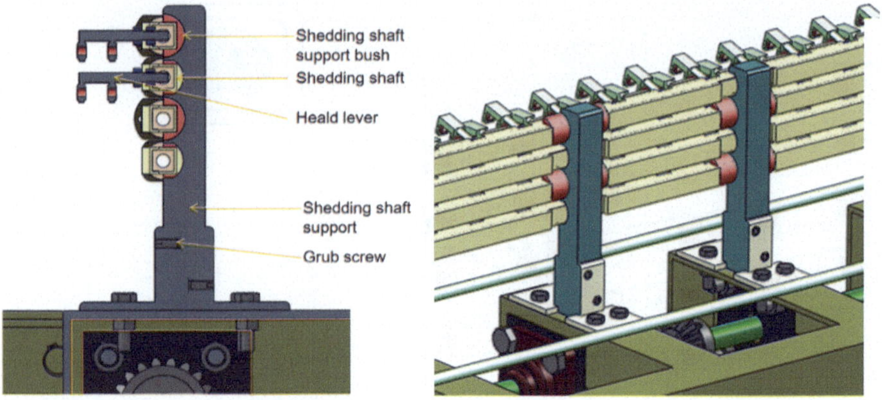

Fig. 7 Support for shedding shafts (left) and alternating supports (red colour) to accommodate in the available space

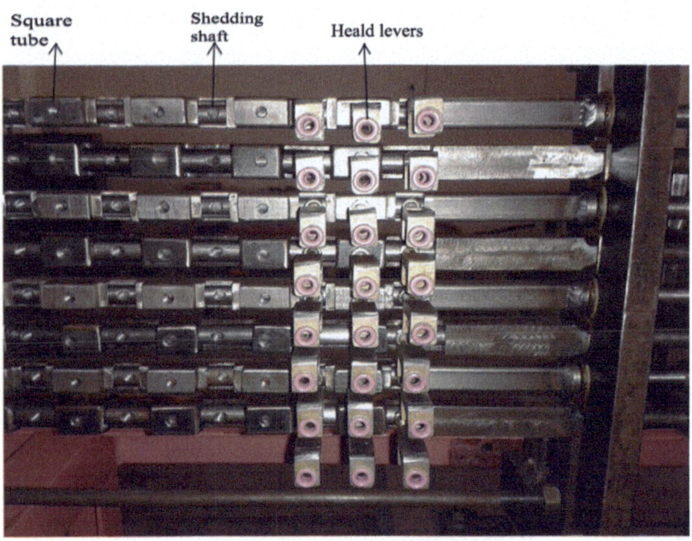

Fig. 8 Tapped holes on the inner and outer shaft for mounting heald levers and the mounted heald levers for few of the tapped holes

shaft are seen through the slots of tube. The Heald-levers are mounted using these tapped holes as shown in Fig. 8.

The shedding shaft and square tube are mounted on the shedding frames with their end-caps. These shedding frames can slide on the rigid structure mounted on the shedding table as shown in Fig. 9.

The Drive for the shedding system addresses the following requirements:

- Outer shaft and inner shaft rotate in opposite directions to achieve the full shed opening.

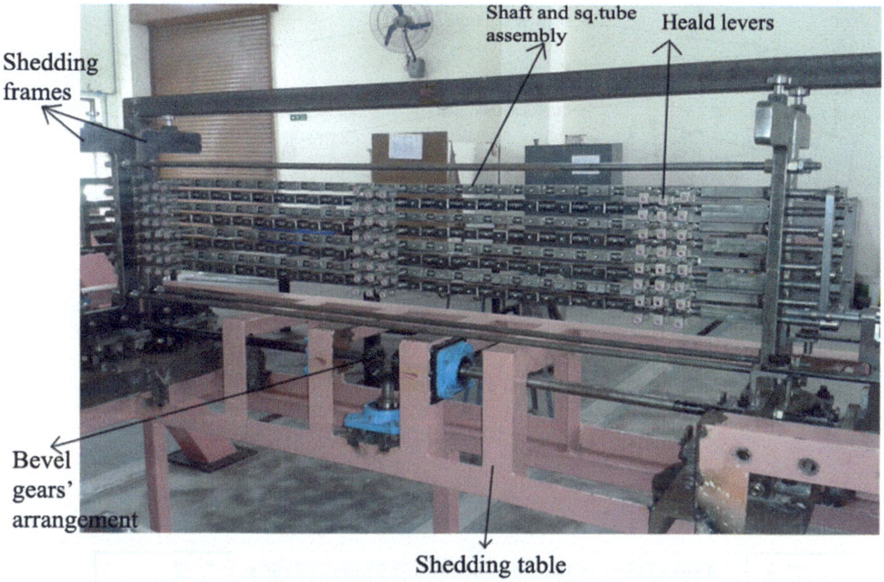

Fig. 9 Shedding table, shedding shaft and frame assembly

- The shafts are driven by single prime mover.
- Translatory motion of inner shaft for vertical shedding.

Two separate mechanisms are adopted for vertical shedding and horizontal shedding. For vertical shedding (shed formation in the horizontal plane), the link arrangements are adopted at both left and right side of the shedding-grid controlling inner shaft and outer tube separately as shown in Fig. 10, adopting the four-bar mechanism. A set of bevel gears are used (Fig. 11) for reversing the direction of motion to move the adjacent yarns in the opposite vertical directions to form the required shed. Figure 12 shows the vertical shedding in operation.

For horizontal shedding another four-bar mechanism is adopted by moving alternative shedding assemblies in opposite directions to form the required shed as shown in Fig. 13. Both the vertical and horizontal shedding are achieved manually by using the hand levers.

3.3 *Picking System*

The requirements of the picking system are as follows:

Shuttle thickness to be less then 20 mm as shed space itself is 25 mm
Distance between consecutive shuttle holders should not exceed 25 mm

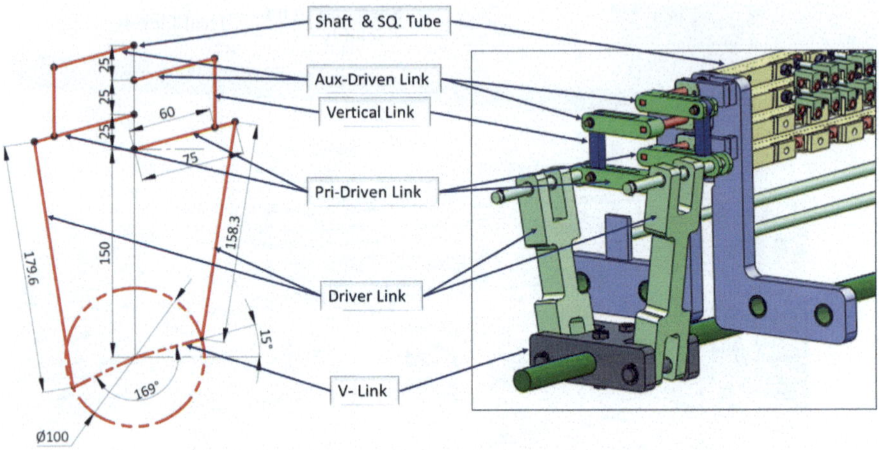

Fig. 10 Four bar mechanism for vertical shedding

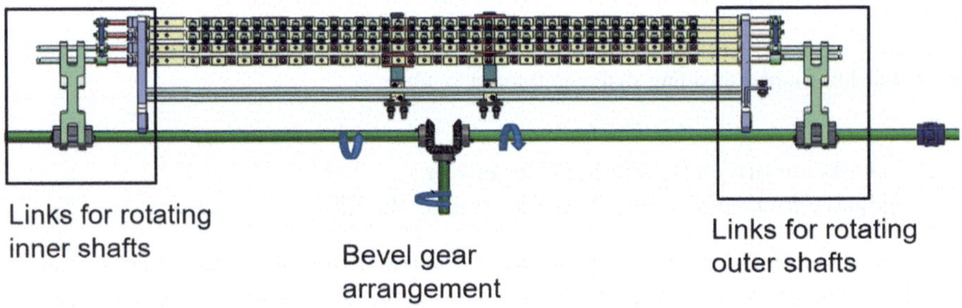

Fig. 11 Bevel gear arrangement to achieve the movement of levers in opposite direction for the outer and inner shafts

Due to converging warp, the shuttles should enter at the precise angle for the respective shed (refer Fig. 14).

To meet the above constraints, vacuum-based picking system was considered. The series of picking pipes, which will also act as vacuum pipes, are arranged at the four sides of the shedding-grid as shown in Fig. 15. The horizontal pipe is divided into two halves, to reduce the un supported length, which will make them more stiffer. This will enable for ease of alignment with other pipe, as can be seen in the figure. To enable to fix the shuttle at the required angle and to suit the angle of the converging warps, the shape of grips of the shuttle is made in circular cross section. This will enable rotating the shuttle to the required angle. O rings have been provided to make the vacuum leak tight.

As these pipes are also acting as vacuum pipes, all the joints were made leak tight with a provision to assemble and disassemble. For this, swage lock fittings were employed. Since the space available between the consecutive pipes is 25 mm only, the joints are staggered to the top and bottom of the pipe as shown in Fig. 16.

Fig. 12 Photograph showing vertical shedding in operation

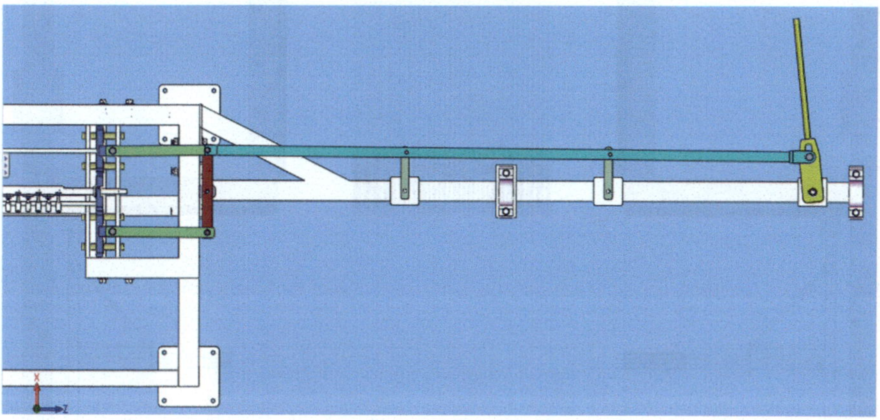

Fig. 13 Four bar mechanism for horizontal shedding

The square tube acts as vacuum manifold for the top section of the vertical picking arrangement. Similar manifolds are provided on all the four sides. Two types of joints are proposed, one is O ring-based and other swage lock-based vacuum tight joints. The tubes that are taken through the manifolds and the tubes which are connected on the top side have been provided holes in the zone of manifold as shown in Fig. 17.

These vacuum pipes are capable of gripping the shuttles. These picking pipes are evacuated using the vacuum reservoir. When the vacuum valve is open, the pipes will grip the shuttle and when the vacuum valve is closed and vent valve is open, the pipes will release the shuttles. This system is adopted to transfer the shuttles horizontally and vertically on LM rails manually through handle wheels. By moving the picking pipes close to each other and by using the vacuum, the shuttles can be transferred

Fig. 14 Angular disposition of shuttles in the sheds

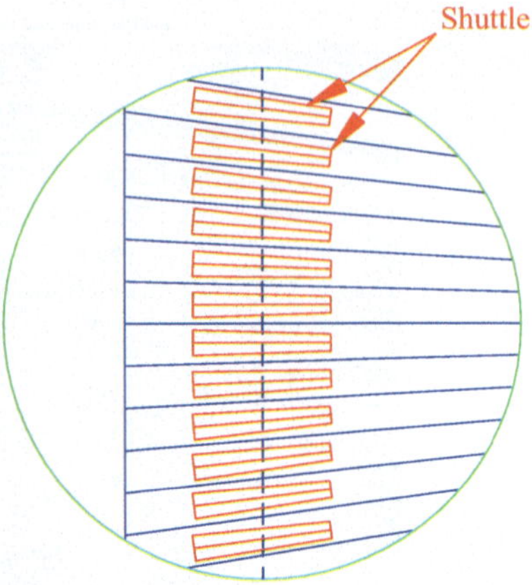

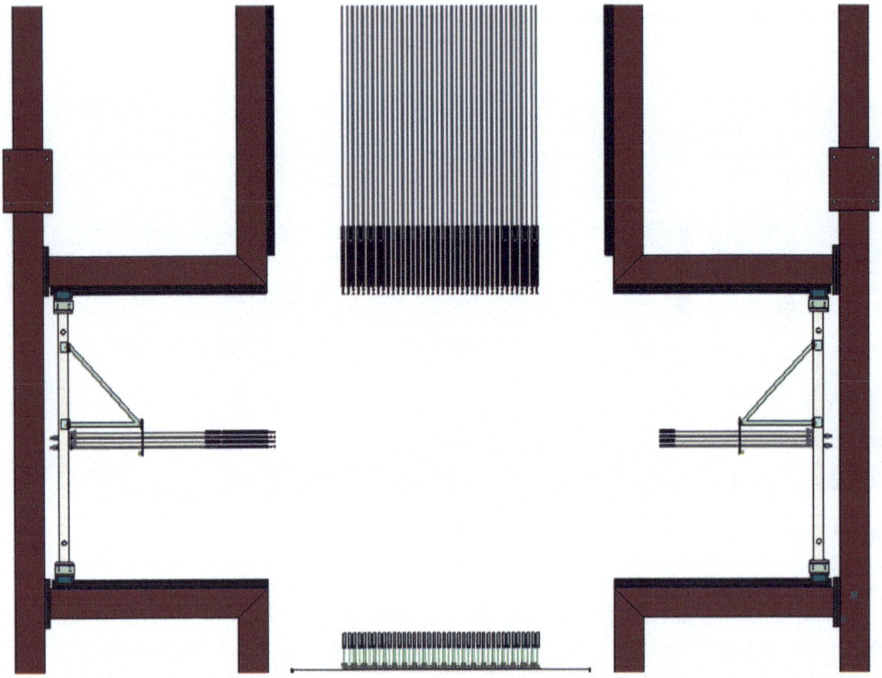

Fig. 15 CAD model of picking system (both horizontal and vertical picking)

to the other side to insert the weft yarn. It is to be noted that no manual interference is required to move the shuttles (in the shuttle zone) except for moving the handle wheels which is away from the picking zone.

Figure 18 indicates the driving arrangement for the vertical motion of picking pipes. The weight of the moving arrangement is balanced by dead weights through suitable pulley arrangement. The power is transmitted manually using hand wheel.

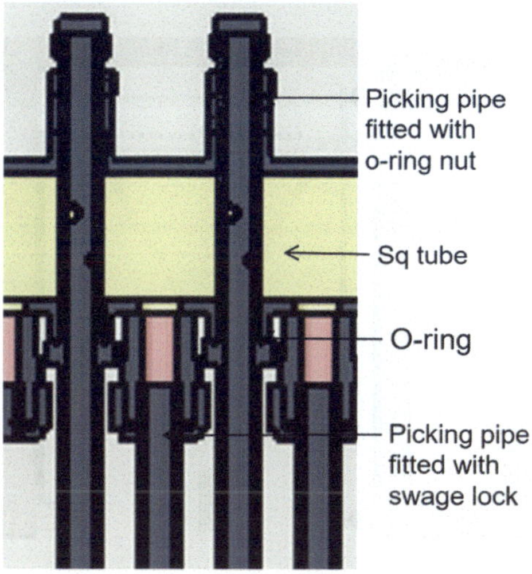

Fig. 16 Alternating O ring and Swage lock approach for picking pipes

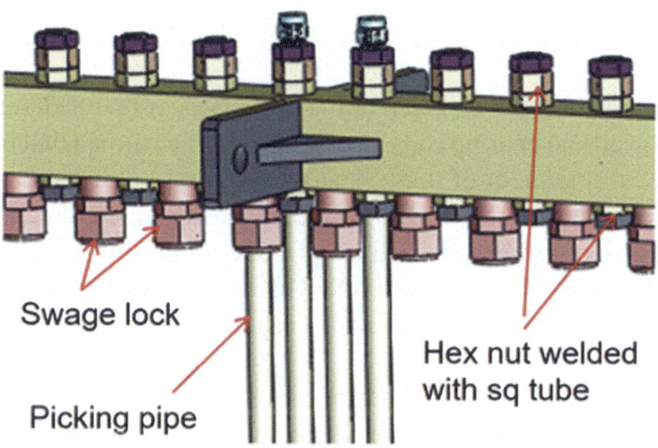

Fig. 17 Staggering of swage lock fittings in constrained space and Vacuum piping sealing arrangement for picking pipes

The hand wheel drives the timing belt connected on both ends of the vacuum manifold, which holds the shuttle tubes. The dead weight is adjustable to match the weight and it is guided as shown.

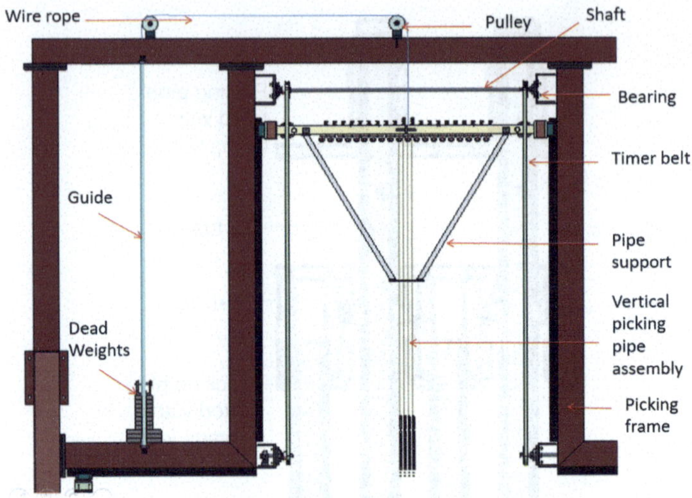

Fig. 18 Driving arrangement for picking assembly

3.4 Beat-Up System

Figure 19 shows the conceptual sketch of the primary beat-up mechanism. It consists of horizontal and vertical beat-up rods having good surface finish that move in respective guides. The arrangement of these beat-up rods and guides are such that, when the beat-up rods move towards the fell of the cloth, the square area formed by both horizontal and vertical guide rods will keep reducing proportionally like a cone to match the reduced cross section at the fell. The reduction is from 1000 to 120 mm. Accordingly, the guide rods are positioned at required angles to achieve the required converging motion.

Figure 20 shows the CAD model arrangement indicating beat up and guide rods positions. Two numbers of pneumatic cylinders with the required stroke length have been positioned along the top guide rods as shown in Fig. 21. Using these pneumatic cylinders and suitable directional control valve, the primary beat-up motions with both forward and reverse movements an be achieved automatically.

3.5 Take-Up System

The movement of the take up has to be synchronized with that of beat up, in order to maintain uniform tension of warps.

To achieve this, it is proposed to use the motive force and instantaneous position of beat-up guides, used for both primary and secondary beat-up operations. All the warp yarns are fixed to the perforated sheet, having same number of holes to that of number of warp yarns drilled in a grid of 120 × 120 mm. The perforated sheet is

Fig. 19 Top and side view of the primary beat up

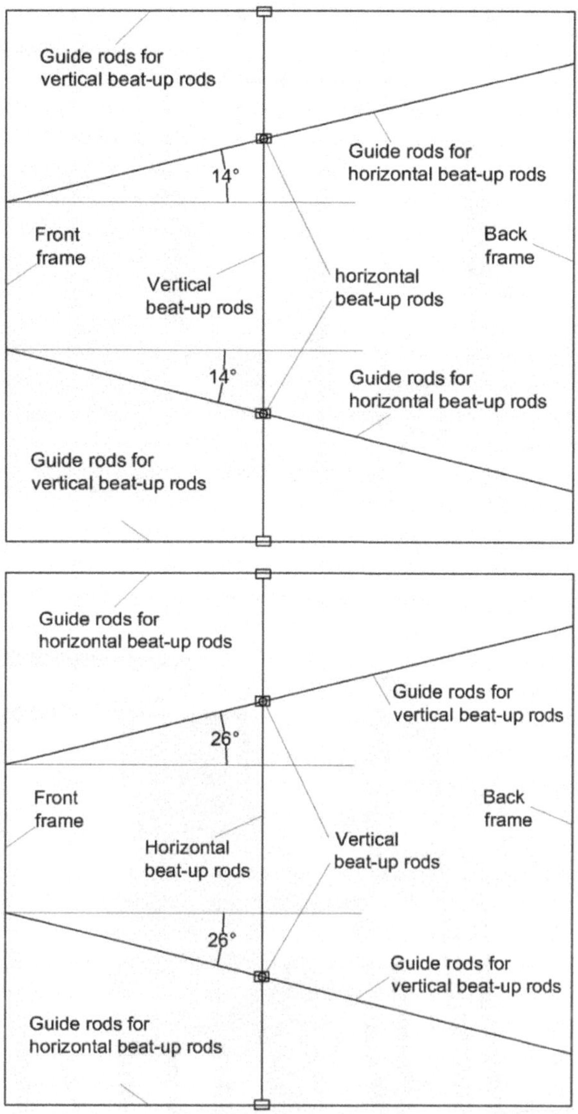

fixed to the take up bed as shown in Fig. 22. Take up bed is driven by lead screw. The lead screw rotation will pull back the perforated sheet, thus the take-up happens for the preform.

The lead screw has to be rotated to the required angle of rotation, based on the pitch of the screw, to get the required preform production rate. Assuming that the preform has to be moved back 1 mm for every cycle of weaving, we can obtain the required angle of rotation based on the pitch of the screw selected.

When lead screw of 25 mm pitch is rotated by 360°, the take up bed will move back by 25 mm. To move 1 mm, the lead screw has to be rotated by 360/25 = 14.4°.

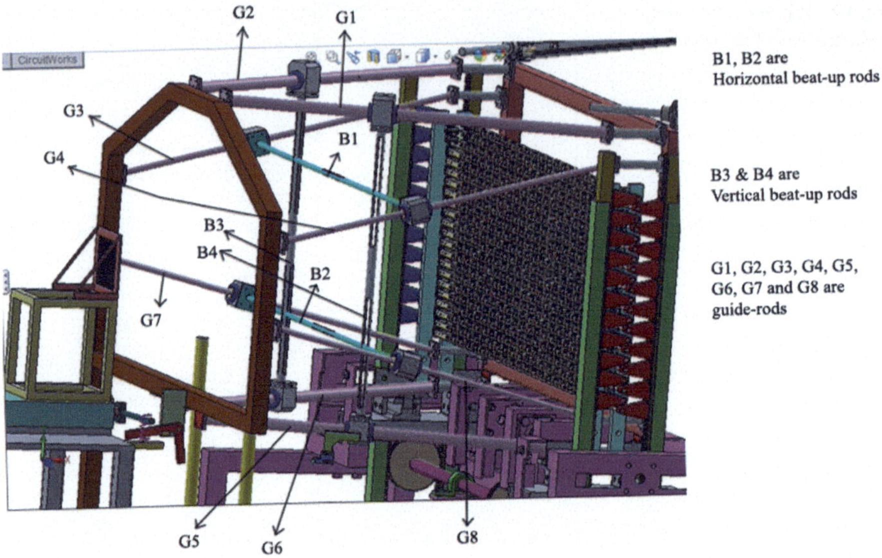

Fig. 20 CAD model of primary beat up

Fig. 21 Pneumatic cylinders for actuation of beat-up mechanism

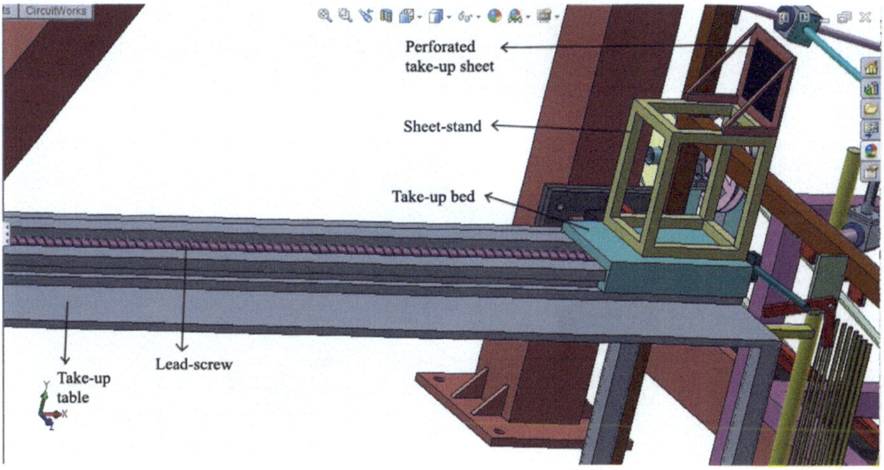

Fig. 22 CAD model of take up arrangement

The motion of beat-up rods is tapped through a series of links as shown in Fig. 42, to rotate the lead screw by the required angle. To ensure that lead screw rotates only in one direction, a ratchet spanner is fixed to the lead screw, which transfers motion only in one required direction. The ratchet spanner is fixed to the link of appropriate length (Fig. 23).

To tap the motion from the beat-up mechanism, pusher links are provided which transmits the motion from moving the pillow block of beat-up mechanism to the

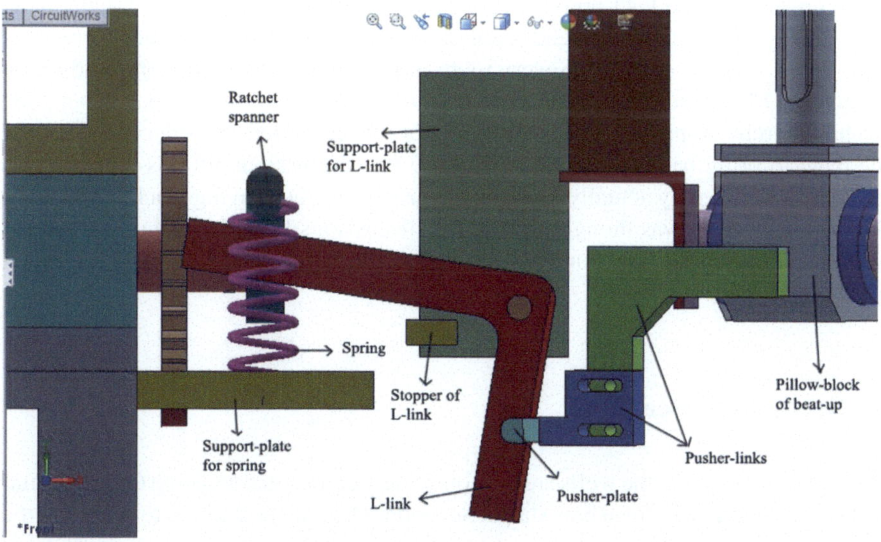

Fig. 23 CAD model of pusher links (in design stage)

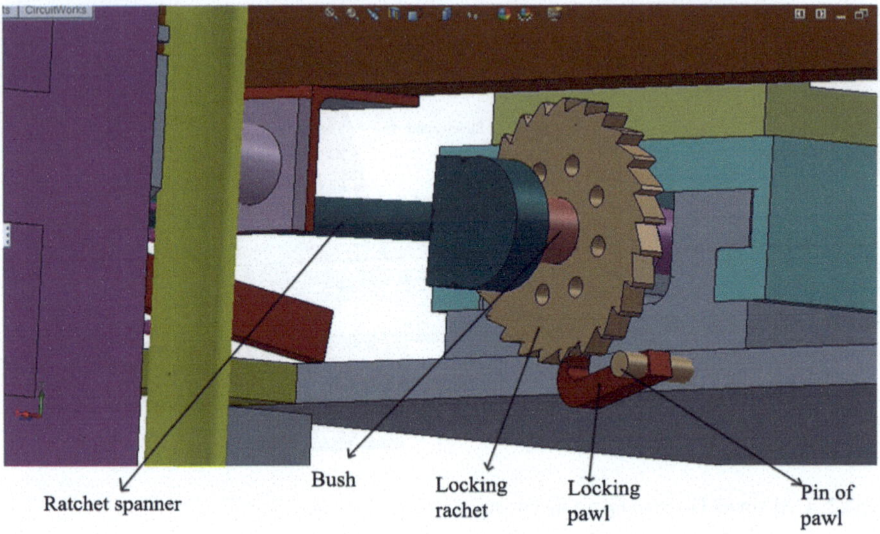

Fig. 24 Ratchet pawl mechanism to lock the take-up lead screw

ratchet spanner. L shaped link with suitable spring is provided to convert the reciprocating motion of the beat-up rod to the oscillating motion. This oscillating motion is transmitted to the lead screw through the ratchet spanner, which transmits the force only in the required direction. When the spanner is lifted, then there will not be any take-up, but when the spring pulls the spanner, lead screw will rotate and take-up happens.

To ensure that the lead screw is rotated only to the required angle, ratchet pawl mechanism is provided to the lead screw as shown in Fig. 24. The ratchet is provided with 25 no. of teeth so that for every tooth movement of the ratchet, the lead screw rotates by 14.4°, there by pulling the take up bed by 1 mm. The pawl locks the ratchet, thereby stopping further movement of the lead screw. Spring is provided on the pawl to bring back to its initial position. The discussions till this juncture was practically realised by actually building the prototype 3D weaving machine. Moving forward, upscaling was thought of, and the discussions from this point, is confined to the drawing board. The concepts have been thoroughly evaluated and presented here.

4 Secondary Beat Up

In the primary beat up, the wefts are not completely compacted and both the horizontal and vertical wefts will form bow shape. Secondary beat up is required to complete the compaction of the 3D preform. Several challenges are there in arriving at the suitable mechanism to complete the beat up. Some of them are space limitation, insertion and retraction of the reeds in between several yarns, which are closely placed in all the

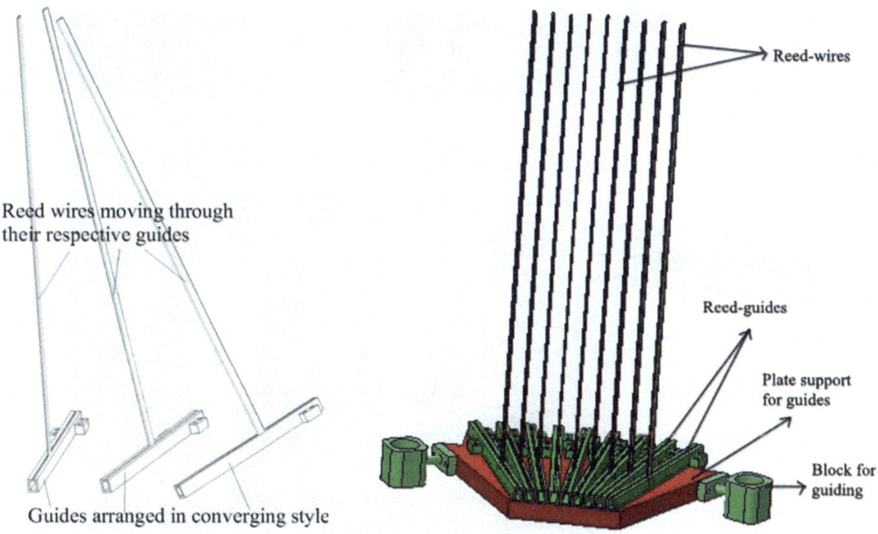

Fig. 25 Reed wires with guides for secondary beat up and the arrangement of reed wires with guiding block

three directions., i.e., x, y and z, ensuring complete compaction, applying sufficient force for compaction, fast retrieving of reeds after beat up in a pre-determined path, etc.,

In this concept, individual reed wires moving in respective guides are provided that will make them to move towards the fell in the required path. Every reed wire will have a predefined path. To meet this requirement, every reed wire is provided with guides in a converging style as shown in Fig. 25.

As secondary beat up is close to cloth fell, and space between them is very minimal, we do not need as many reed wires as number of warps. It is proposed to provide one reed wire for every 4 warp threads. In this case, about 9 reed wires with respective guides are proposed.

These 9 reed wires are fixed on to the plate, so that all can be deployed or inserted by lifting through guides for secondary beat up, as shown in Fig. 25. Either pneumatic cylinder or simple rack and pinion arrangement can be provided to lift and lower the secondary beat-up plate.

Further to retract the reed wires after beat up, every reed wire guide is provided with springs. The reed wires are pushed by the primary beat-up rods, which will provide sufficient force through pneumatic cylinders. The CAD model of the fixing arrangement of secondary beat up in 3D weaving machine is shown in Fig. 26.

One more concept of secondary beat up is explained which is likely to be more effective to reduce the bow effect. A frame is provided to guide the reed wires at desired positions, which include both deployment and removal. Also, the use of a mechanism to aid support for the reed wires on the existing beat-up rods to avoid a support structure for the reed wires. This is accomplished by designing a system that combines the motion of primary beat up with the elements of secondary beat up. The primary beat-up system acted as a support structure for the elements of secondary

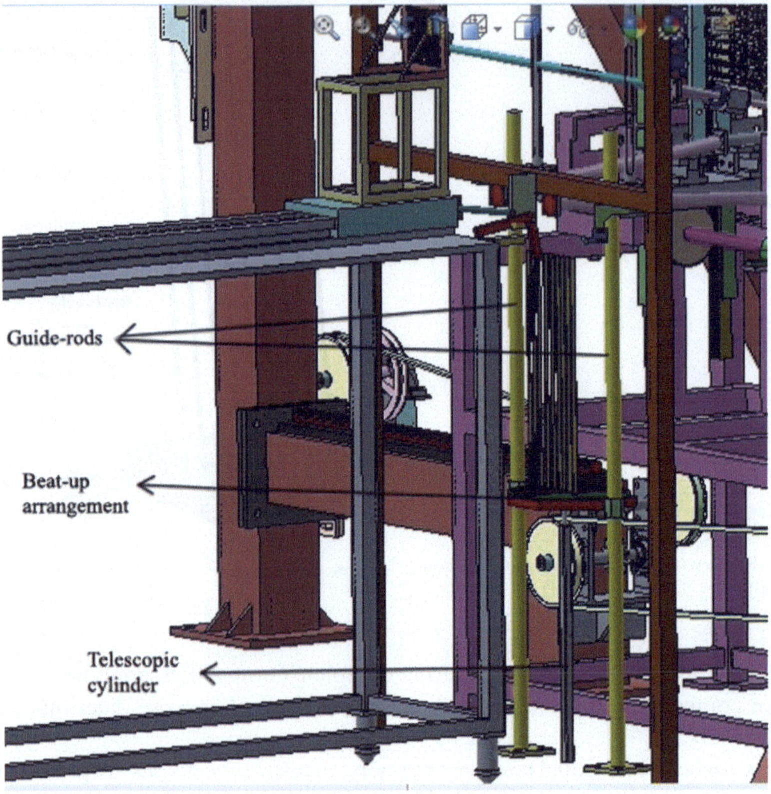

Fig. 26 CAD model of fixing arrangement of secondary beat-up in 3D weaving machine

beat up. The motion of this primary beat up enabled motion for the secondary beat-up elements without additional mechanisms.

Intricate mechanisms of the secondary beat-up elements ensured localized beat up of both Y and Z-weft layers. Minimal energy requirement due to the self supporting nature of secondary beat-up elements reduced the requirement of additional actuations. Combining all the features, the effect of formation of a bow in the preform was negated.

Fig. 27 indicates alternate concept of the secondary beating setup.

The prototype development was carried out for 8 × 40 viz., 8 shedding elements, 8 picking elements in the horizontal plane and 40 picking elements in the vertical plane. However design was carried for 22 × 40. The following section discusses these details.

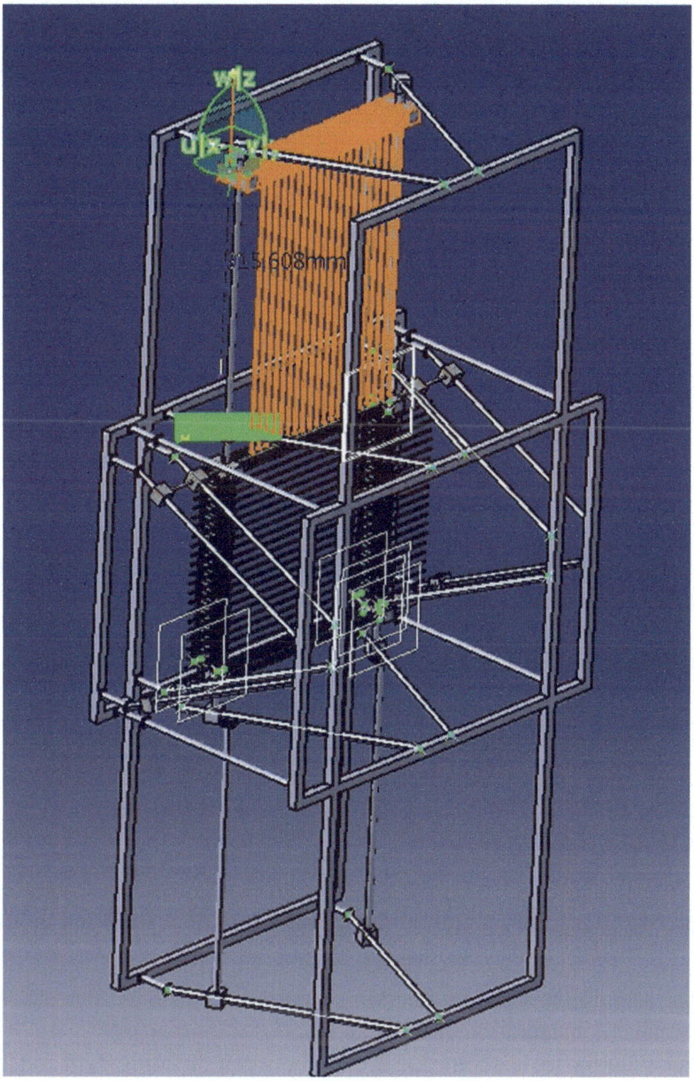

Fig. 27 Secondary beat-up arrangement with complete frame

4.1 Alternate Approach for Shedding Mechanism

In the event there is an increase in the number of shedding shafts the force required for operation of the shedding system cannot be taken by the links, and therefore, an alternate approach as shown in Fig. 28 (either full gears or segmental gears) can be used. In case the full gear system is used, it will call for change in shed space of 25 mm which is not preferred. However, the partial gear concept can be adopted

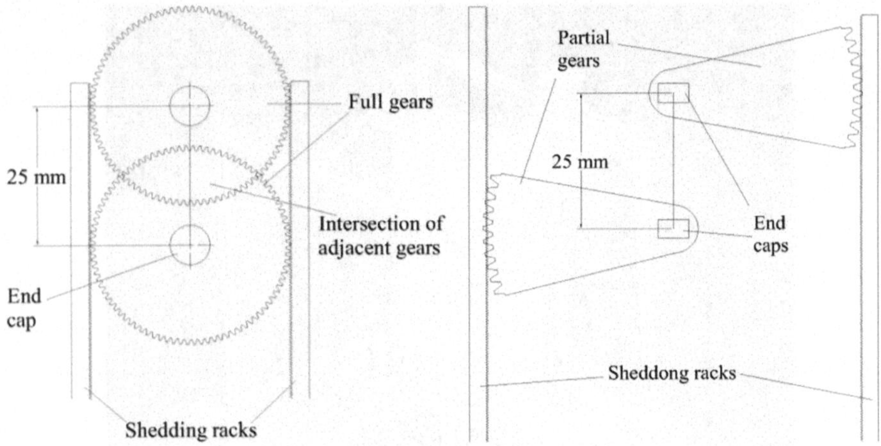

Fig. 28 Shedding using full size gears or segmental gears

without any change in the shed geometry. The other advantage of using partial gears is that, higher the diameter of partial gear, higher the torque that can be transmitted.

Figure 29 indicates spur gear concept to actuate racks for moving shedding racks in opposite directions. Figures 30 and 31 illustrate the CAD model for 22 shaft assembly and the motorized gear assembly for driving the shafts.

4.2 *Design Details for Sizing of Various Components for Vertical Shedding*

To estimate the total torque required for actuating vertical shedding elements, the following factors have to be considered:

(a) Torque required to rotate the shedding shafts along with its 20 heald-levers and 20 threads (consideration of thread tension is important).
(b) Torque required to overcome the friction between shedding shaft and the square tube.
(c) Torque required to hold the racks and to move those racks against gravity.
(d) Torque required to overcome the friction between racks and their support.

The sum of all the above values will be the torque required on one side of shedding assembly. For two sides of the shedding assembly, the torque required would double. The calculations for the above 4 cases are shown below.

(a) *Torque required to rotate the shedding-shafts along with its 20 healdlevers and 20 threads (consideration of thread tension is important).*

The torque required for any rotating body is calculated as, $M_t = I \times \alpha$

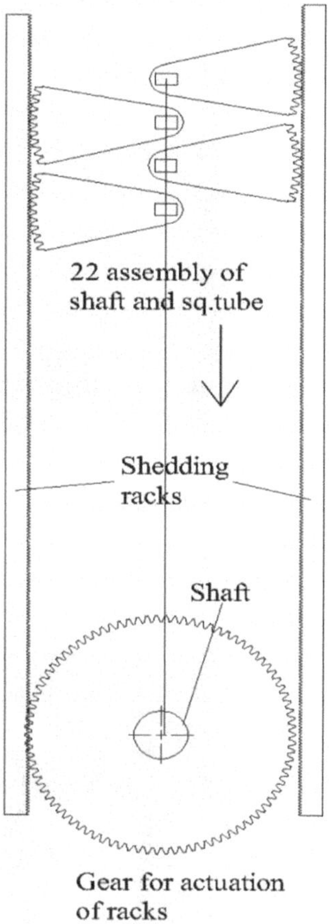

Fig. 29 Spur gear concept for actuating the racks

where, I = Mass-moment of inertia of that body.

$\propto$ = Angular acceleration of the rotation, adopting $\propto = 0.1$ rad/s^2.

The mass-moment of inertia (M.I) of different components is listed below.

M.I of single shedding-rod = 2.09×10^{-5} kg-m^2.

(Including its end cap).

M.I of 20 heald-levers = 1.05×10^{-3} kg-m^2.

M.I of single partial gear = 1.25×10^{-4} kg-m^2.

Total of all the above, M.I = 1.2×10^{-3} kg-m^2.

Then, torque required by these considerations = $1.2 \times 10^{-3} \times 0.1 = 1.2 \times 10^{-4}$ N − = 1.2×10^{-1} N-mm.

Now, considering 20 threads of the shedding shaft.

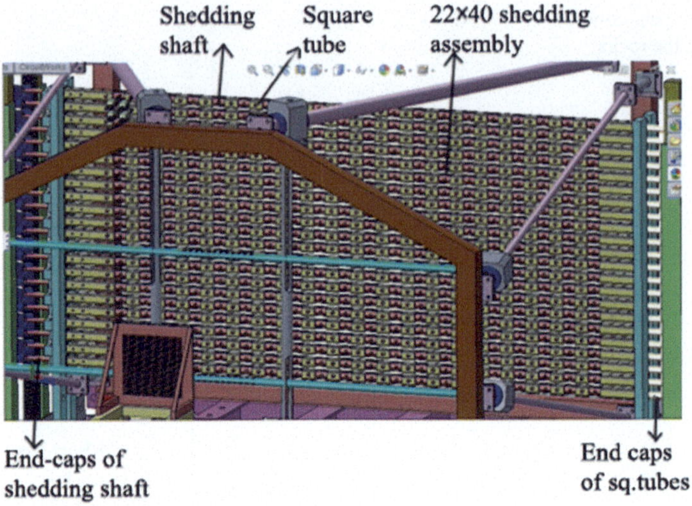

Fig. 30 CAD model of 22 shaft assembly

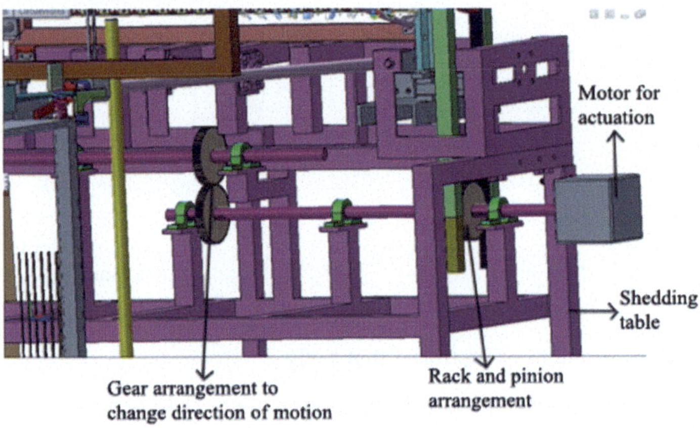

Fig. 31 Motorized gear assembly for shedding motion

Each of the thread is with tension 30gm = 0.03 kg.

Total thread tension on the shedding shaft = 0.03 × 20 = 0.6 kg.

The weight applied on the shedding shaft, W = m × g = 0.6 × 9.81 = 5.88 N.

Torque required to hold the threads in their position = W × d = 5.88 × 70 = 412.2 N-mm.

Force required to move the threads against gravity, F = m × a where, a = 7×10^{-3} m/s^2

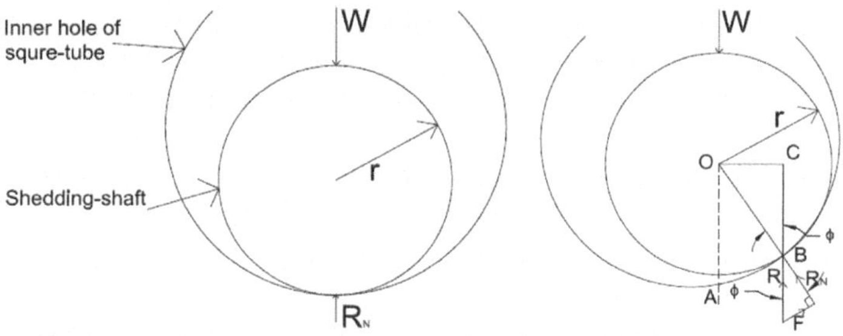

Fig. 32 Friction between shedding shaft and support at initial position and after a small rotation

$$F = 0.6 \times 7 \times 10^{-3} = 4.2 \times 10^{-3} N$$

Torque required to move the threads against gravity $= F \times d = 4.2 \times 10^{-3} \times 70 = 0.3$ N-mm.

Sum of the torques for holding and moving the threads $= 412.2 + 0.3 = 412.5$ N-mm.

Thus the sum of torques of M.I and thread tensions for the single shedding shaft.

$$[M_t]_{1shaft} = 1.2 \times 10^{-1} + 412.5 = 412.62\, N - mm.$$

The torque required for 22 shedding shafts.

$$[M_t]_{22shaft} = 412.62 \times 22 = 9077.64\, N - mm.$$

(b) *Torque required to overcome friction between shedding-shaft and support.*

Figure 32 shows the initial and final conditions of shedding shaft at the support. After initial movement, the reaction point changes from 'A' to 'B' and the quantity of reaction also changes from 'R_N' to 'R'. The load 'W' and the distance 'OC' creates the torque, $T = W \times OC$, which is the minimum torque required to overcome from the friction between two surfaces.

$$T = W \times OB \times \sin \varphi = W \times r \times \sin \varphi.$$

$T = W \times r \times \tan \varphi = W \times r \times \mu$; $\tan \varphi = \sin \varphi$ since φ is very small and $\mu = \tan \varphi$.

In our case, $W =$ weight of the shedding shaft $= 1 \times 9.81 = 9.81$ N.

$r =$ Radius of the shaft $= 6 \times 10^{-3}$ m.

$\mu =$ Co-efficient of friction between two surfaces $= 0.1$

Then, torque required for one shedding shaft to overcome from friction.

$$[M_t]_{1shaft} = 9.81 \times 6 \times 10^{-3} \times 0.1 = 5.88 \times 10^{-3} N - mm$$

The torque required for 22 shedding shafts to overcome the friction.

$$[\mathrm{Mt}]_{22\mathrm{shafts}} = 5.88 \times 22 = 129.36\,\mathrm{N-mm}$$

(iii) *Torque required to hold the racks and to move those racks against the gravity.*

Selecting the suitable dimensions of the rack as below

$$\text{Length} = 1000\,\text{mm, Face width} = 40\,\text{mm, Depth} = 21\,\text{mm}$$

Then, volume of the rack = 1000 × 40 × 21 = 840,000 × 10^{-9} m^3.

Then, mass of the rack = 840,000 × 10^{-9} × 7850 = 6.6 kg.

Weight of the rack = 6.6 × 9.81 = 64.7 N.

Let us consider that the diameter of the gear is 130 mm and radius is 65 mm.

Torque required to hold the rack against the gravity = 64.7 × 65 = 4205.5 N-mm.

Force required to move this rack against the gravity

$$\mathrm{F} = \mathrm{m} \times \mathrm{a};\ \text{where, a} = \mathrm{r} \times \alpha = 65 \times 0.1 = 6.5\,\mathrm{mm/s^2}$$

$$\mathrm{F} = 6.6 \times 6.5 \times 10^{-3} = 0.0429\,\mathrm{N}$$

Torque required to move the rack against gravity = 65 × 0.0429 = 2.7885 N-mm.

Then, total torque required to hold the rack and move it against gravity = 4205.5 + 2.788 = 4208.29 N-mm.

Torque required for both the racks will be 4208.29 × 2 = 8416.5 N-mm.

The sum of torques of concepts ‘a’,’ b’ and ‘c ‘is = 9077.64 + 129.36 + 8416.5 = 17,623.5 N-mm.

Consider service factor = 2.

Total torque required = 17,623.5 × 2 = 35,247.15 N-mm.

(iv) *Torque required to overcome from the friction between racks and their support.*

Figure 33 shows generation of friction during the motion of rack through its guides.

Now, considering friction between rack and its support, there are mainly 3 types of forces to be observed.

F_t = Tangential force applied by the gear on the rack.

F_r = Radial force applied by the gear on the rack.

Fig. 33 Friction between rack and its support

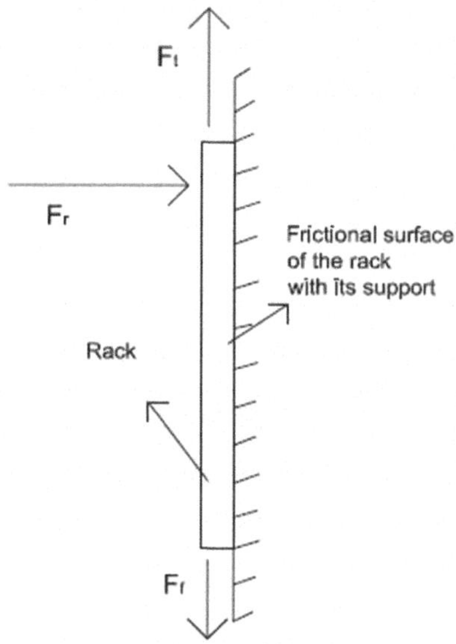

F_f = Frictional force exerted between rack and support during the motion.

Tangential force, $F_t = \frac{M_t}{r_g}$

M_t is torque applied by the gear.

r_g is radius of the gear.

$$F_t = \frac{35247.15}{65} = 542.26\,N$$

$$F_r = 542.26 \times \tan 20^\circ = 197.36\,N$$

Frictional force generated, $F_f = \mu \times F_r$; considering $\mu = 0.2$

$$F_f = 0.2 \times 197.36 = 39.47\,N$$

Force required to overcome the friction between rack and support = F_t = 542.26 + 39.47 = 581.73 N.

Torque for one side of shedding assembly = 581.73 × 65 = 37,812.68 N-mm.

Same amount of torque is required to rotate the shedding rods along with heald levers and the threads at the opposite side of the shedding assembly, to rotate the square tubes along with heald levers and the threads.

Then, the total torque required = 37,812.68 × 2 = 75,625.36 N-mm.

But by considering unpredictable factors and adopting the service factor as 2, the total torque required for the complete shedding = 75,625.36 × 2 = 151,251 N-mm.

Design of all machine elements like gears, shafts and keys of the vertical shedding will be based on this 151251 N-mm torque.

Design of shedding gears

Steel-C45-(untreated material) may be used for the gears.

Allowable static stress, $\sigma_o = 233$ MPa,

Young's modulus of elasticity, $E = 2 \times 10^5$ MPa.

Assuming Number of teeth on gear, z = 90.

Torque to be transmitted by gear, $M_t = 151{,}251$ N-mm.

The modulus of the gear can be calculated from Lewis equation

$$m^3 = \frac{2M_t}{\lambda\pi^2 z y c_v \sigma_0}$$

where m = modulus of the gear, M_t = Torque to be transmitted by gear.
$\lambda = \frac{b}{\pi m}$ is constant and generally considered as $\lambda = 4\pi$.

$$y = 0.54 - \frac{0.912}{z} = 0.154 - \frac{0.912}{90} = 0.144$$

$C_V = \frac{3}{3+v_m}$; v_m is the pitch line velocity.

$$V_m = \frac{\pi m n}{60 \times 1000} = \frac{3.142 \times m \times 90 \times 2}{60 \times 1000} = 0.0095 \times m \text{ in m/s.}$$

Then, $C_v = \frac{3}{3+0.0095m}$
Then,

$$m^3 = \frac{(2 \times 151251)\,(3 + 0.0095m)}{4 \times \pi^2 \times 90 \times 0.144 \times 233 \times 3}$$

By solving the above equation for modulus, m = 1.4. Select standard module 1.5 Then,
Diameter of the gear, d = m × z = 1.5 × 90 = 135 mm.
Face width of the gear, b = 4πm = 4 × 3.142 × 1.5 = 18.852 mm.

$$V_m = 0.0095 \times 1.5 = 0.01425 \text{ m/s};\ C_v = 0.09955$$

select width b = 20 mm.

Allowable stress, $\sigma_{all} = \sigma_o \times C_v = 233 \times 0.9955 = 232$ MP.
Induced stress 'σ_{ind}' is calculated as below.
$\sigma_{ind} = \frac{(2\times 151251)}{4\times\pi^2\times 90\times 0.144\times 0.955\times 1.4^3} = 216.38$ MPa.

Checking the safety of the gear

Dynamic load on the gear, $F_d = \frac{F_t+21v(F_t+bc)}{21v+\sqrt{(F_t+bc)}}$

where, Tangential force, $F_t = \frac{M_t}{r_g} = \frac{151251}{67.5} = 2241$ N
Pitch line velocity, v = 0.01425 m/s; Face width, b = 20 mm.
Dynamic factor, C = 1144; Selected from DDHB, assuming for 0.1 mm error of gear.
Then,

$$F_d = \frac{2241 + 21 \times 0.01425 \times (2241 + 20 \times 1144)}{(21 \times 0.01425) + \sqrt{2241} + 20 + 1144)}$$

$F_d = 58.89$ N.

Wear load, $F_w = d \times b \times Q \times K$.

Where, teeth ratio factor, $Q = \frac{2Z_2}{Z_1+Z_2}$; But for rack and pinion arrangement $Z_1 = Z_2 = 90$

Then, Q = 90.

$$K = 10.43 \times \sigma^2_{-1c} \times \sin\alpha/E_o$$

where, surface fatigue strength, $\sigma_{-1c} = (2.75\ H_B – 69)$; H_B is Brinell hardness number = 200.
then, $\sigma_{-1c} = (2.75 \times 200 – 69) = 481$ MPa.
Equivalent young's modulus $E_0 = \frac{2E_1E_2}{E_1+E_2}$. But here $E_1 = E_2$.
Then, $E_o = 2 \times 10^5$ MPa.

$$K = (1.43 \times 481^2 \times \sin 20)/(2 \times 10^5) = 0.5658$$

Then,

$$F_w = 130 \times 20 \times 90 \times 0.5658 = 132,397.2\,N$$

$F_w > F_d$; The design is safe.

Design of shafts

There are 2 shafts 'S1' and 'S2'. The shaft 'S1' is mounted by the gears 'G1' and 'G2'. The shaft 'S2' is mounted by the gear 'G3' and 'G4'. This arrangement is shown in Fig. 34.

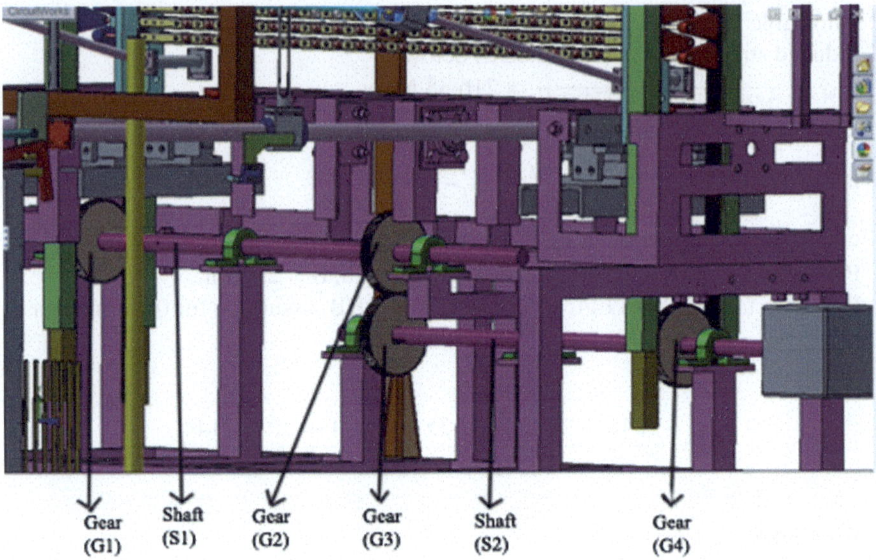

Fig. 34 Gears and shafts used for vertical shedding

The flow of torque should be analyzed during the design of shafts. The torque '151,251 N-mm' is supplied to 'S2' from which '75,625.5 N-mm' is used for the actuation of two racks and another '75,625.5 N-mm' is supplied to the shaft 'S1'.

Thus, it is clear that, 'S2' is under more torque and load when compared with 'S1'. Design 'S2' and then adopting same for 'S1' will be the safer design.

Considering the forces at the position of 'G4'.

The tangential force exerted by the gear 'G4' in vertical directions on each rack

$$\mathrm{F_t} = \frac{M_t}{r_g} = \frac{75625.5/2}{65} = 582\,\mathrm{N}.$$

The '582 N' force exerted by the 'G4' is, one in vertically upward-direction for one rack and another in downward direction for another rack. Thus, the opposite forces cancel out.

The radial force exerted by the gear 'G4' in horizontal direction

$$\mathrm{F_r} = \mathrm{F_t} \times \tan\,20^\circ = 582 \times \tan\,20^\circ = 212\,\mathrm{N}.$$

This case is also similar to that of the tangential force. The two forces of value '212 N' will be applied in horizontal opposite directions on two opposite racks, and hence they also cancel out.

The only considerable force is, the weight of the gear 'G4 and G3', acting vertically downward. Figures 35 and 36 shows the force and bending moment diagram of shaft, respectively.

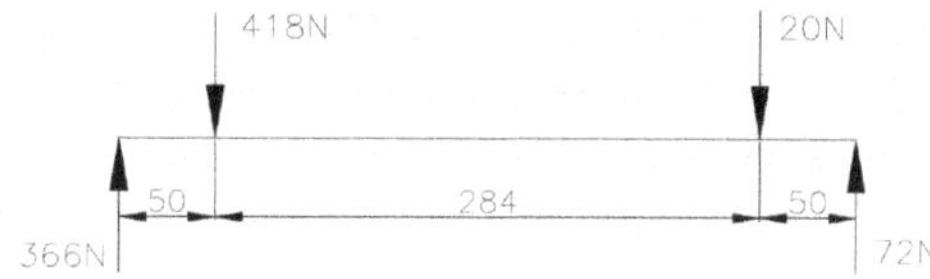

Fig. 35 Vertical force diagram due to gear G4 and G3

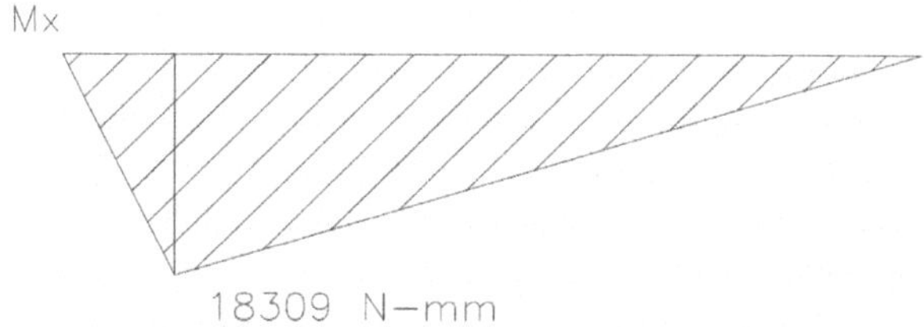

Fig. 36 B.M diagram of shaft of Fig. 35

$W_{G4} = 20$ N.
Now, considering the forces at the position 'G3'.
The tangential force in the horizontal direction

$$F_t = \frac{M_t}{r_g} = \frac{75625.5}{65} = 1163.46\,\text{N}.$$

The radial force in the vertical downward direction,

$$F_r = F_t \times \tan 20 = 582 \times \tan 20 = 212\,\text{N}.$$

Weight of the gear in the vertically downward direction,

$$W_{G3} = 20\,\text{N}$$

Then, total vertical downward force is 398 + 20 = 418 N.
The maximum B.M due to vertical forces, $(M_b)_V = 366.18 \times 50 = 18{,}309$ N-mm.
No horizontal force acting on the shaft at Gear G4 location.
The maximum B.M due to horizontal forces at G4 $(M_b)_H = 0$.
The maximum B.M due to horizontal forces (, $(M_b)_H = 1012 \times 50 = 50{,}600$ N-mm.
The resultant maximum B.M due to vertical and horizontal forces,

$$(M_b) = \left[(M_b)_v^2 + (M_b)_H^2\right]^{1/2} = (50600^2 + 18309^2)^{1/2} = 53,811\,\text{N-mm}.$$

The maximum torsional moment applied, $M_t = 151{,}251$ N-mm.
Then according to the maximum shear stress theory, the diameter of the shaft.

$$D_o = \left\{\frac{16}{\pi\tau_{max}} \times \sqrt{\left[(C_m \times M_b)^2 + (C_t \times M_t)^2\right]}\right\}^{1/3}.$$

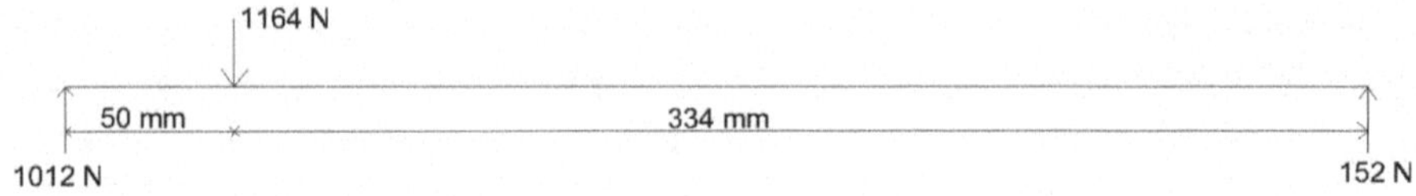

Fig. 37 Horizontal force diagram at G3 location

Adopting suitable values from DDHB,
Shock and fatigue factor for bending, $C_m = 1$.
Shock and fatigue factor for torsion, $C_t = 1$.
Selecting the shaft material, S.A.E 1045 annealed. For this material $\tau_{max} =$ 180 MPa.
By adopting FOS = 2, $\tau_{max} = 90$ MPa.

Then,

$$D_o = \left\{\frac{16}{\pi \times 16} \times \sqrt{\left[(1 \times 53811)^2 + (1 \times 151251)^2\right]}\right\}^{1/3} = 20.24\ \text{mm}.$$

As a standard, select the diameter of the shaft $D_o = 25$ mm.

Similar shaft can be adopted as 'S1', which will be also the safer design. Because, the torsion and B.M applied on 'S1' is smaller when compared with that of the 'S2'.

Figures 37 and 38 show the horizontal force diagram at G3 location and the corresponding BM diagram.

Figure 39 shows the dual crank-based drive assembly for horizontal shedding as against the single crank assembly for load reduction. Also, the dual crank approach simplifies the assembling of the drive. The dual crank assembly houses a rolling joint that moves in the vertical direction between bracket and crank as shown in Fig. 40.

The horizontal shedding system consists of two cranks mounted on the shaft (Fig. 41). Each crank with bracket is connected to the shedding arm by pin joint (Fig. 42). Pillow blocks are provided to support the shaft. The crank-arm connection is very important to produce opposite motions for frames. One shedding arm is connected to crank at its lower circumferential part and another shedding arm is connected to other crank at its upper circumferential part. This type of connection makes both shedding arms to move in opposite directions (while one arm moves

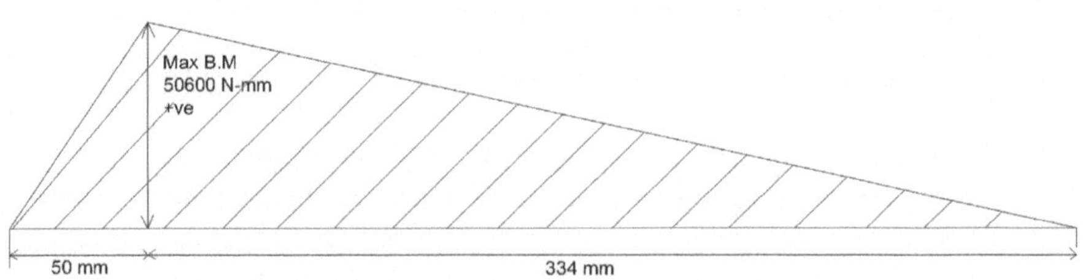

Fig. 38 B.M diagram of shaft of Fig. 36

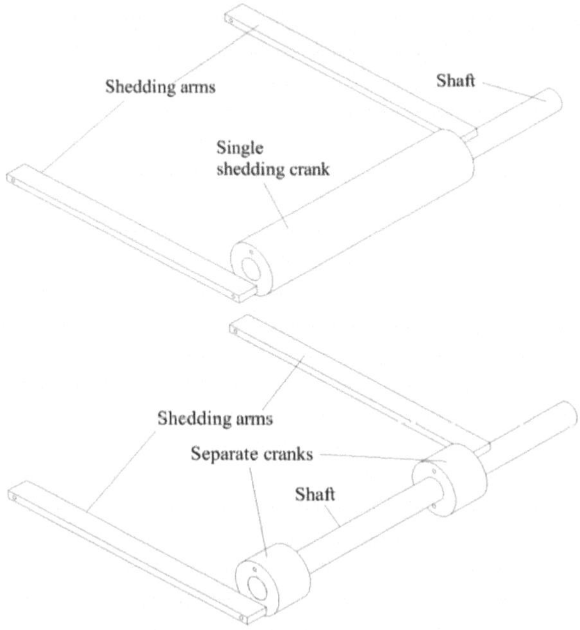

Fig. 39 Dual crank assembly (bottom) as against Single crank (top) for shedding

front the other will move back and vice-versa) for same shaft rotation (Fig. 43). The size of crank and angle of crank rotation determines the availability of 25 mm shed space. The gear motor can be used for rotation of shaft. The selected motor should be capable of rotating the shaft in both clockwise and anticlockwise directions to achieve dual directional shedding in horizontal direction.

Ref:

- Machine Tool Design Handbook by CMTI (Central Machine Tools Institute), India.
- Machine Design Handbook, Second edition, K.Lingaiah, McGraw-Hill Handbooks.
- Fundamentals of Applied Kinematics, D.C. Tao., Addison-Wesley Publishing Company.
- Kinematics and Dynamics of Plane Mechanisms, Jeremy Hirschhorn, McGraw-Hill Book company.

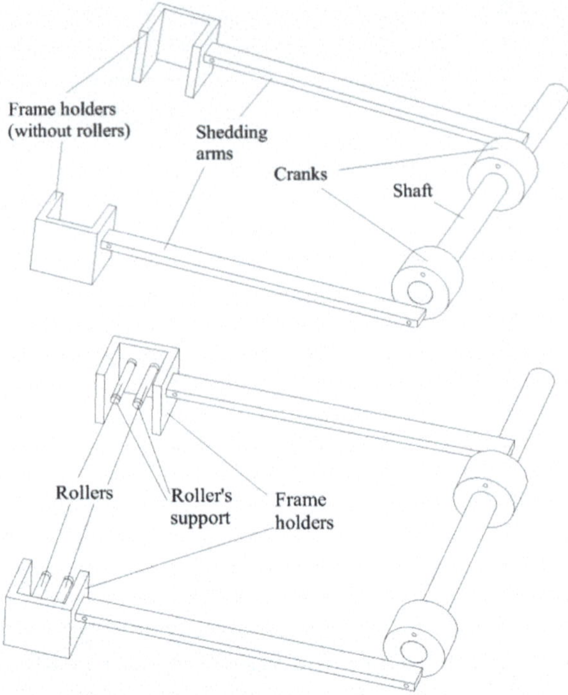

Fig. 40 Frame bracket without (top) and with (bottom) rollers

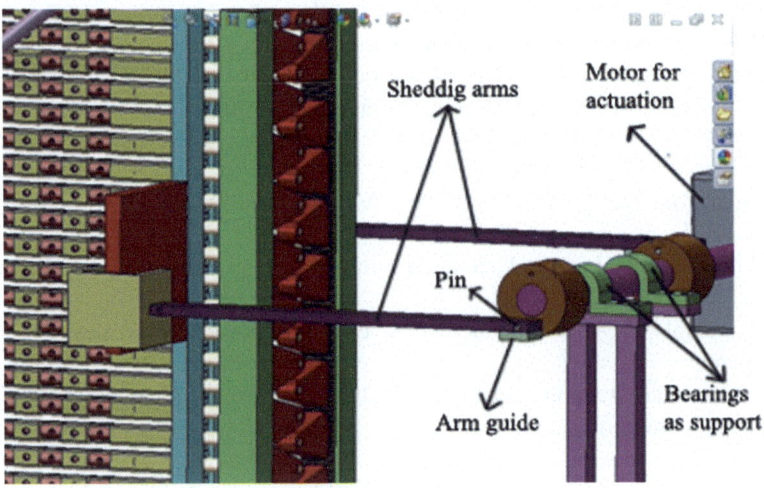

Fig. 41 Horizontal-shedding arms, guides, pins and supports

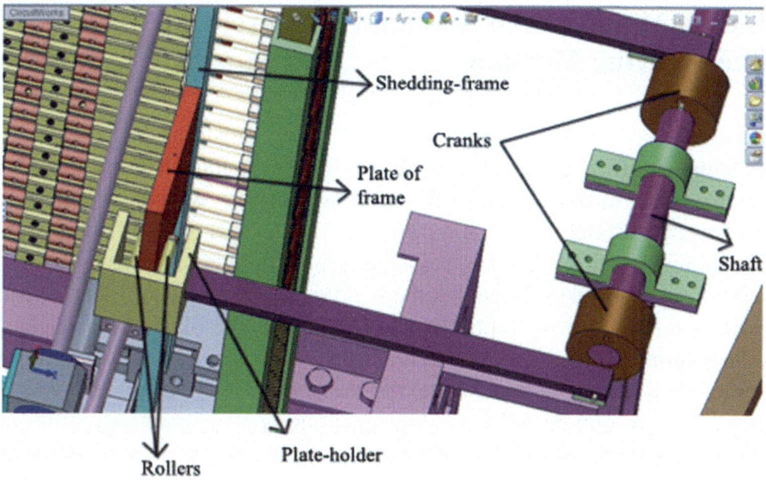

Fig. 42 Horizontal shedding shaft, cranks, rollers and holder

Fig. 43 Arrangement of cranks to achieve opposing linear motions

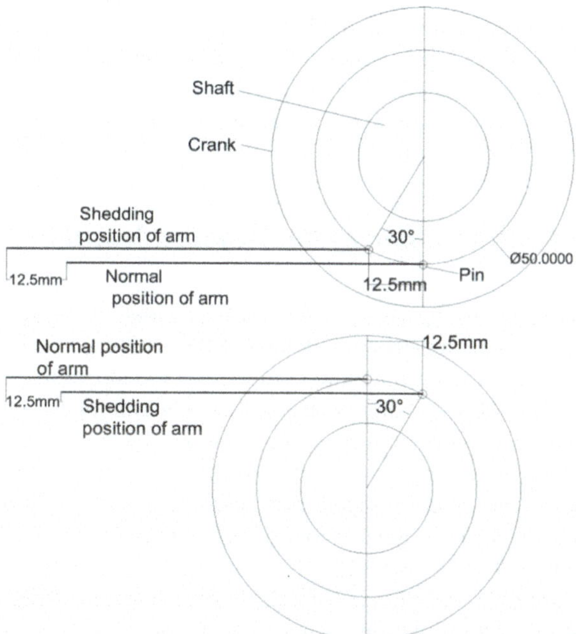

5 Winding Machine: An Auxiliary Requirement to the Weaving Machine

In such unique developments such as the dual plane shedding-based 3D weaving machine, support infrastructure as available in the textile industries may not be suitable to be used. One such example is the winding machine, which is required for the winding of the creel bobbins and the pirn. The type of pirn used is unique, and the available winding machines are not suitable to meet the winding requirements. The envisaged requirements have been that the winding machine is required to uniformly wind textile yarns and most importantly must be capable of winding packages of various dimensions. The other standard features such as tension control arrangement and adaptability to different types of textile yarns have also been addressed. While the design and development of a winding machine is a commonality in the textile industry, the uniqueness of the machine developed here is that traverse speed adjustment is achieved by providing a cone drum drive arrangement wherein changing the position of the belt on the cone-drum drive assembly varies the speed as per required pattern.

Also, the traverse length adjustment is achieved by providing a double rocker arm mechanism. The double rocker arm mechanism comprises of two rocker arms, wherein sliding the connecting pin of one rocker arm in the slot provided changes the traverse length. Most importantly this machine design is completely mechanical based without the involvement of any form of electronics.

5.1 *Description of the Winding Machine*

Referring to Figs. 44, 45, 46, 47, 48, 49 and 50 the following paragraph provides the mechanical construction details of the winding machine.

The winding machine comprises an electric motor (1) mounted on the bottom portion of a rectangular frame (2). This motor is connected to the main shaft (6) positioned horizontally above the motor by means of pulleys (3 & 5) and a belt (4). The main shaft is mounted onto the said rectangular frame using pillow blocks (7 & 8), and it is provided with additional pulleys (9 & 10). The pulley (9) on the main shaft is connected to the horizontally mounted cone drum assembly through a belt (11) and pulley (12). The cone drum assembly consists of a pair of driver cone drum (13) and driven cone drum (14) positioned one below the other and tapered in opposite directions. These are mounted on the said frame using square flange units (15 & 16). The cone drums are connected to each other by a belt (17) guided through guide plates (18). The guide plates are mounted on a screwed rod (20) in the horizontal plane by means of a bracket (19). The screwed rod is provided with a handle (21) and is mounted onto the said frame using a pillow block (22) and a square flange unit (23). The driven cone drum is connected to the worm shaft (26) at the right angle in the horizontal plane through a pair of bevel gears (24 & 25).

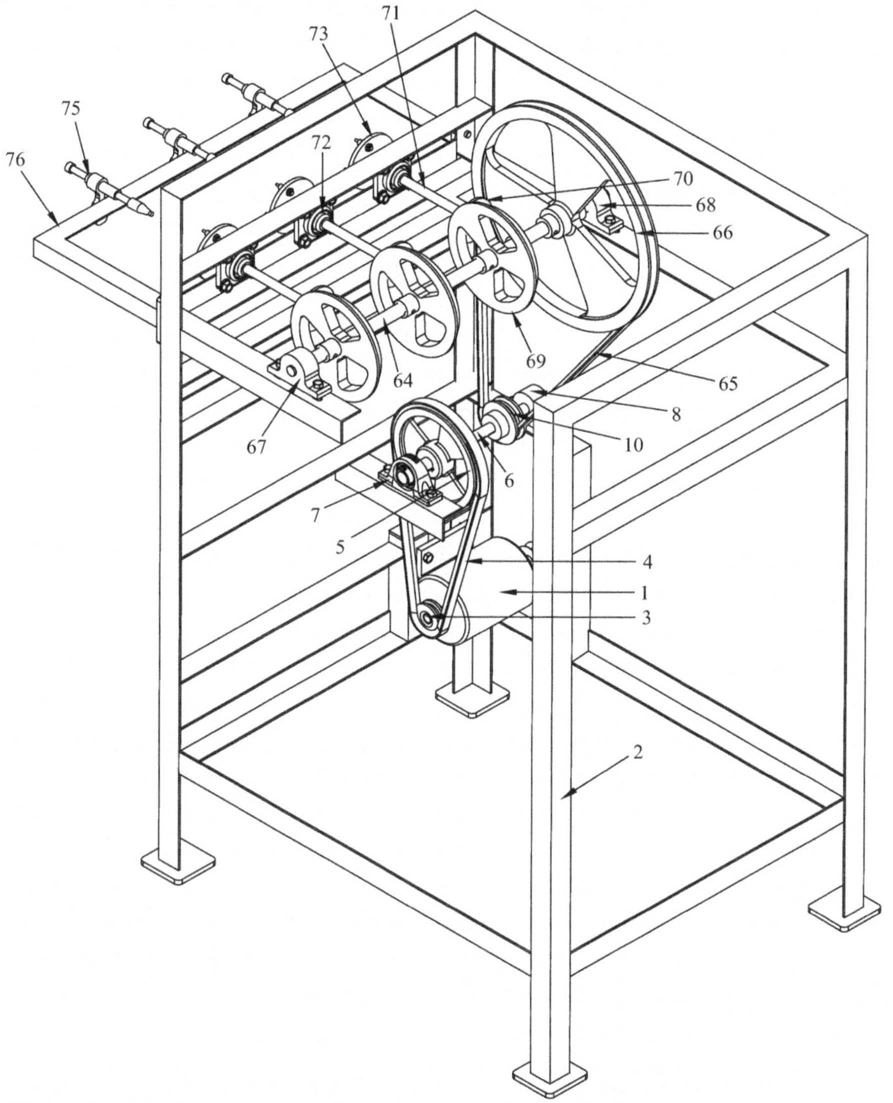

Fig. 44 Isometric view of the bobbin drive mechanism. Parts list: 1. Electric Motor, 2, 76 Rectangular frame, 3, 5, 66. Pulleys, 4, 65 Belt, 6. Main shaft, 7, 8, 67, 68. Pillow blocks, 10. Pulley, 64. Intermediate shaft, 69, 70. Friction Pulley, 71. Spindle shafts 72. Square flange unit, 73. Receiving package holder, 75. Spring-loaded clamp

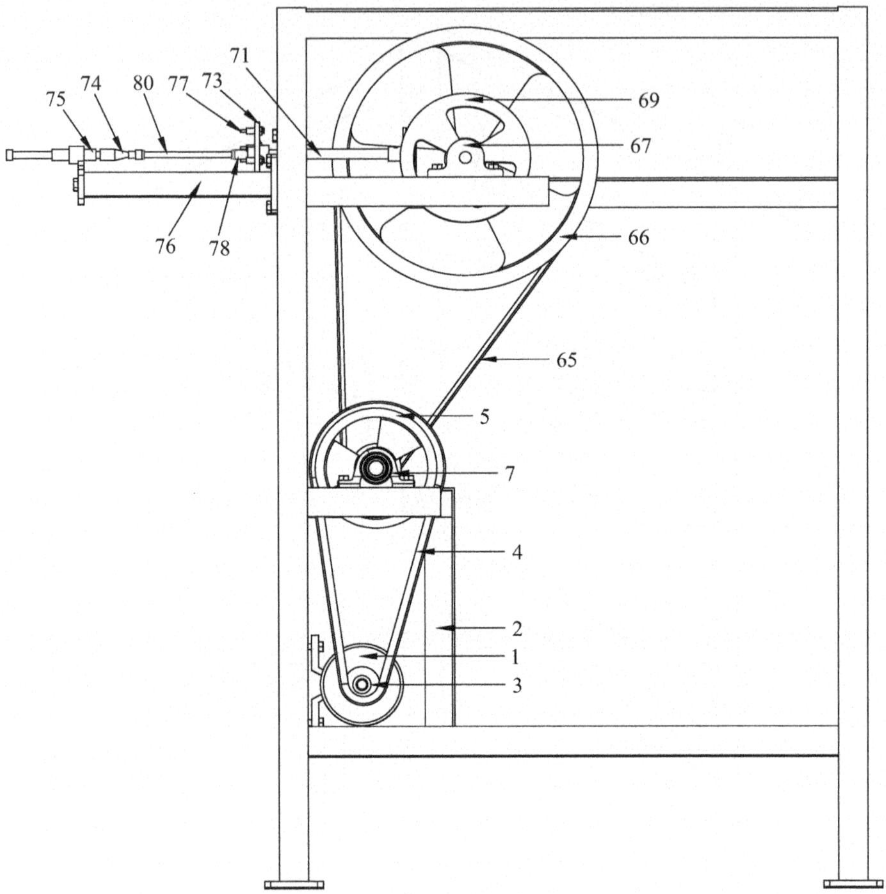

Fig. 45 Side view of the bobbin drive mechanism. Parts list: 1. Electric Motor, 2, 76 Rectangular frame, 3, 5, 66. Pulleys, 4, 65 Belt, 7, 67. Pillow blocks, 10. Pulley, 69. Friction Pulley, 71. Spindle shafts, 73. Receiving package holder, 74, 77, 78. Adaptor 75. Spring-loaded clamp, 76. Frame, 80. Receiving packages

The worm shaft is mounted on the said frame using pillow blocks (27 & 28) and connected to a camshaft (31) through a worm (29) and gear wheel (30). The camshaft consists of a cam (32) followed by a follower (33) mounted on a pin (34) housed in a bracket (35). The follower is connected to a deadweight (37) through a wire rope (36) to ensure the continuous contact of the follower with the periphery of the cam in the horizontal plane. The follower bracket is connected to the double rocker arm (40 & 41) and is positioned in the vertical plane through link rods (38 & 39). The link rods are positioned at right angles to each other in the horizontal plane. An input rocker arm (40) is pivoted to the said frame with the help of a bolt nut (42), which also acts as a fulcrum. The rocker arms are hinged to each other by means of a pin (43) mounted at the top of the input rocker arm, which slides in the slot (44) of the

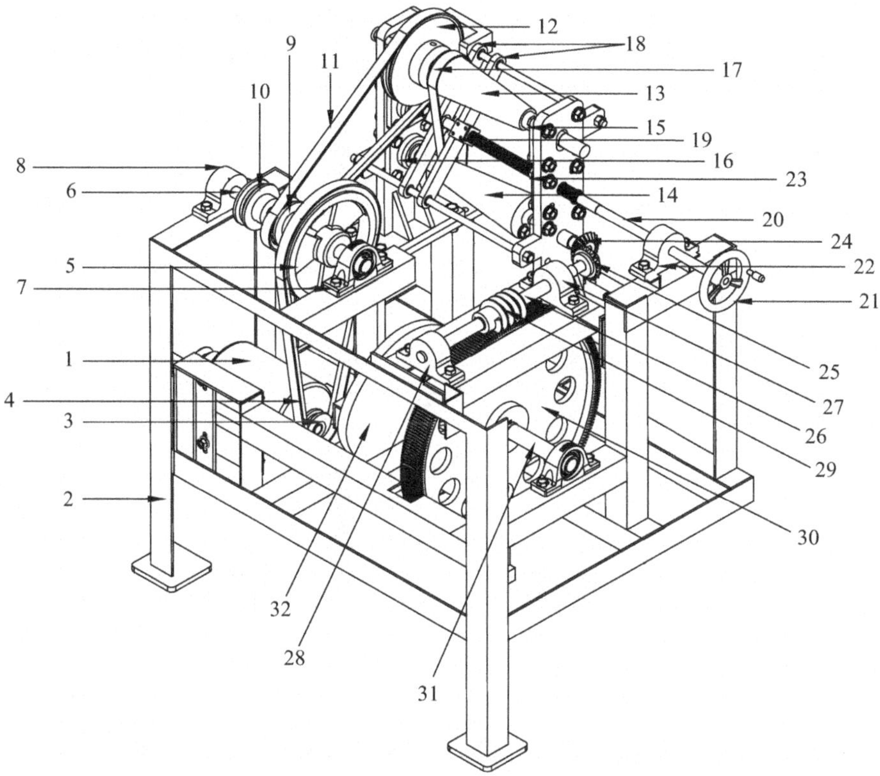

Fig. 46 Isometric view of the cone-drum drive assembly. Parts list: 1. Electric motor, 2. Rectangular Frame, 3, 5, 9, 10, 12 Pulleys, 4, 11, 17. Belts, 6. Main shaft, 7, 8, 22, 27, 28. Pillow blocks, 13. Driver cone drum, 14. Driven cone drum, 15, 16, 23. Square flange units, 18. Guide plates, 19. Bracket, 20. Screwed handle, 21. Handle, 24, 25. Bevel gears, 26. Worm Shaft, 29. Worm, 30. Gear wheel, 31. camshaft, 32. Cam

output rocker arm (41). The bottom end of the said output rocker arm is connected to a guide (47) through link rods (45 & 46) at right angles to each other. The guides are having threaded end caps (48 & 49) mounted on a lead screw (50) in the vertical plane. The lead screw is provided with a handle (51) on the top end and is mounted on the frame using square flange units (52 & 53). The top end of the said output rocker arm is connected to a traverse unit through a link rod (54) and knuckle joint (55) in the horizontal plane. The traverse unit comprises a traverse plate (56) which is positioned at right angles to the said link rod. The traverse plate is connected to the plurality of thread guide assemblies in the vertical plane by means of perpendicular links (57 & 59) and is bolted to said traverse plate through a guide bush (58) mounted on the frame. The thread guide (63) of the said thread guide assemblies is attached to a holder (62) and mounted on a screwed rod (61) by means of a lock nut (60). Another pulley (10) of the said main shaft is connected to the intermediate shaft (64) by means of pulley (66) and belt (65). The intermediate shaft is mounted on the

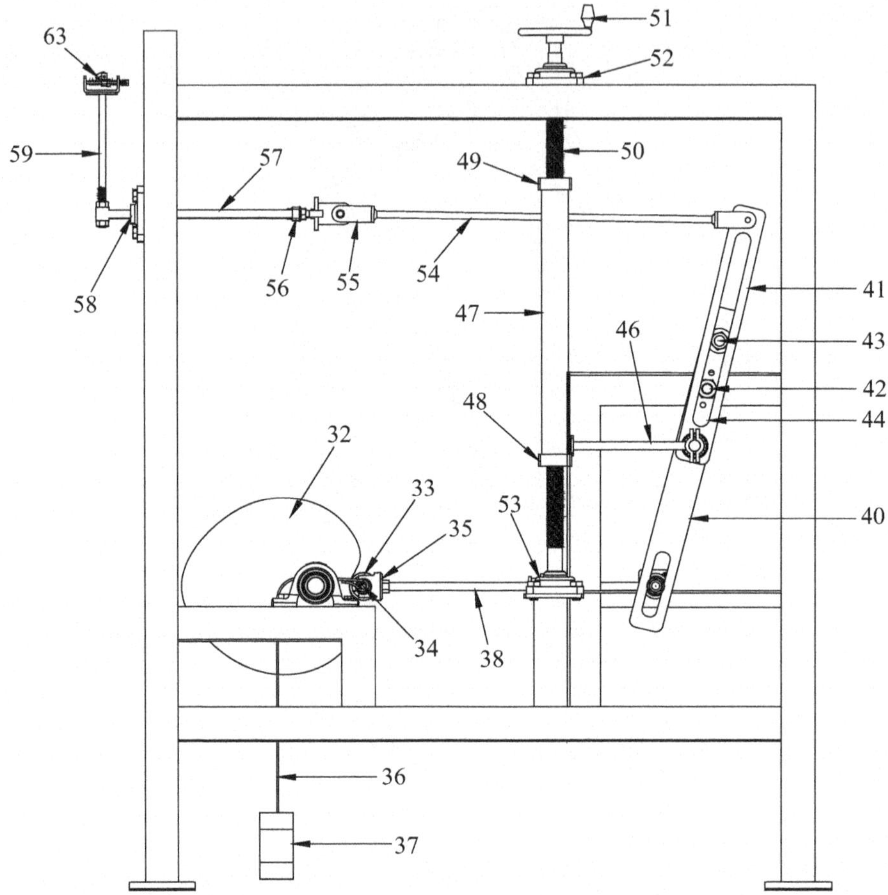

Fig. 47 Side view of the double rocker arm mechanism. Parts list: 32. Cam, 33. Follower, 34. Pin, 35. Bracket, 36. Wire rope, 37. Deadweight, 38, 45, 46, 54. Link rods, 40, 41. Rocker arms, 42. Bolt Nut, 43. Pin, 44. Slot, 48, 49. Threaded end caps, 50. Lead screw, 51. Handle, 52, 53. Square flange units, 55. Knuckle Joint, 56. Traverse plate, 57, 59. Perpendicular links, 58. Guide bush, 63. Thread guide

top portion of the said rectangular frame by means of pillow blocks (67 & 68). The intermediate shaft consists of a plurality of friction pulleys (69) connected to another set of friction pulleys (70) on one to one basis mounted on one end of spindle shafts (71). The other end of each said spindle shaft extends out of the said frame through a square flange unit (72) mounted on the said frame and houses a receiving package holder (73). The receiving package holder holds the receiving package bobbin (79) or pirn (80). The other end of said receiving package is held in position by means of a spring-loaded clamp (75) and suitable adapter (74). The spring-loaded clamp is mounted on the extended portion of the frame (76). The spring-loaded clamp and the receiving package holder are provided with specific adapters (77 & 78) for specific

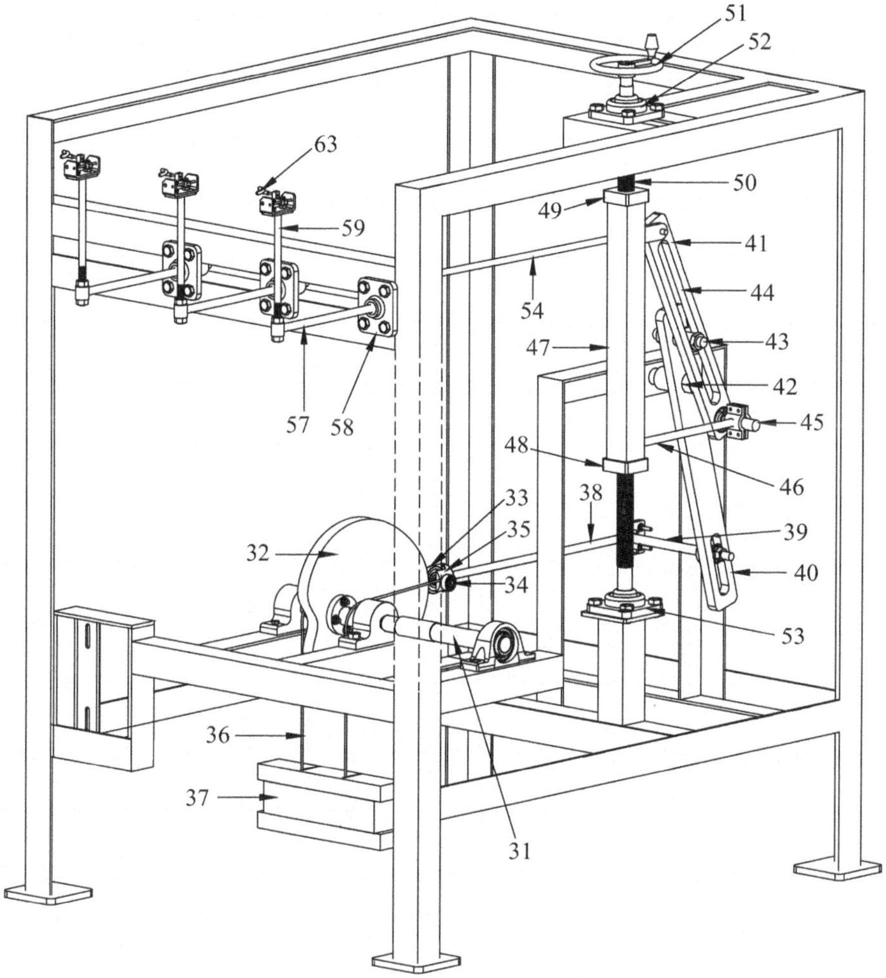

Fig. 48 Isometric view of the thread guide assembly. Parts list: 31. camshaft, 32. Cam, 33. Follower, 34. Pin, 35. Bracket, 36. Wire rope, 37. Deadweight, 38, 39, 45, 46, 54. Link rods, 40, 41. Rocker arms, 42. Bolt Nut, 43. Pin, 44. Slot, 47. Guide, 48, 49. Threaded end cap, 50. Lead screw, 51. Handle, 52. Square flange unit, 57, 59. Perpendicular links, 58. Guide bush, 63. Thread guide

types of receiving packages (79 & 80). The receiving package receives tow (81) through a spring-loaded disc type tensioner (82) from a supply package holder (83) mounted on the bottom of the said frame. The supply package (84) is mounted onto said supply package holder.

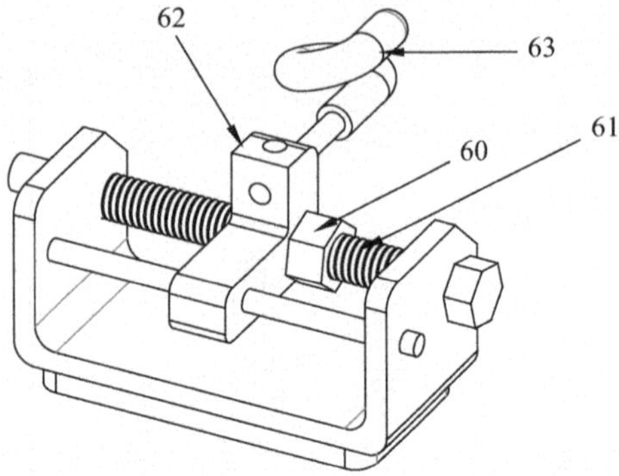

Fig. 49 Isometric view of the thread guide. Parts list: 60. Locknut, 61. Screwed rod, 62. Holder, 63. Thread guide

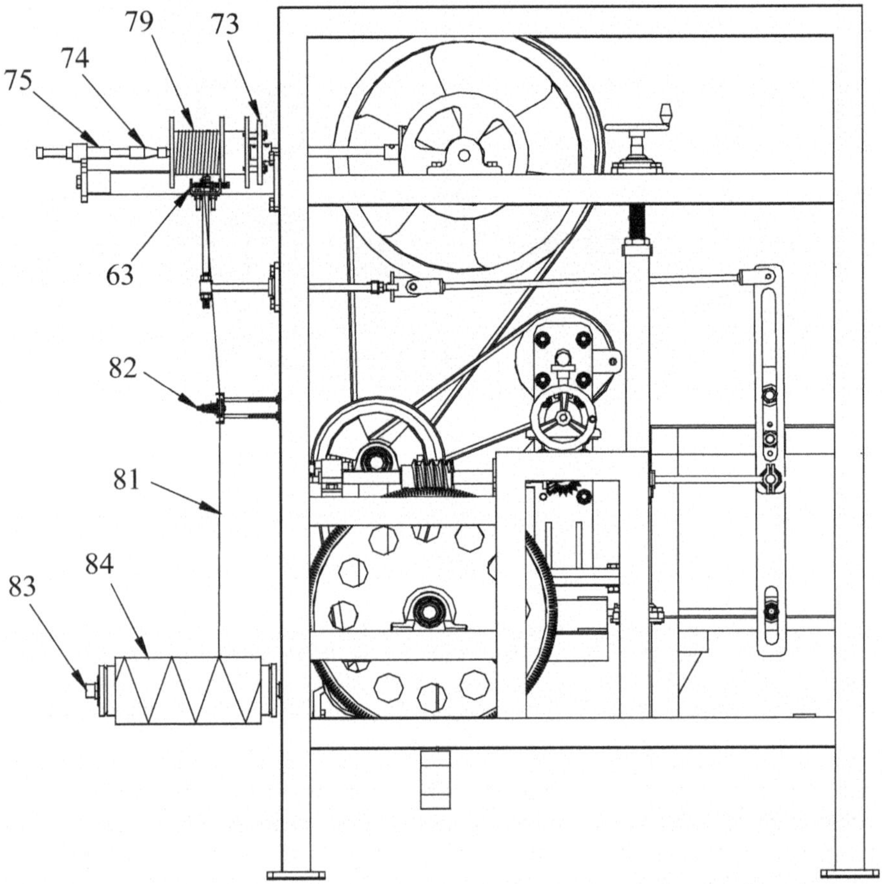

Fig. 50 Yarn path from supply package to receiving package. Parts list: 63. Thread guide, 73. Receiving package holder, 74. Adaptor, 75. Spring-loaded clamp, 79. Receiving package, 81. Tow, 82. Disc type tensioner, 83. Supply package holder, 84. Supply package

5.2 *Working of the Winding Machine*

Now, the supply package mounted on the supply package holder comprises tows passing through a spring-loaded disc type tensioner which is adjustable. The supply package holder is provided with bearings for unwinding with minimum friction. The tow then passes through a thread guide onto a bobbin or pirn. The holders for the bobbin and the pirn are different due to the differences in their dimensions. These holders get the drive from the spindle which transmits the motion to the bobbin or the pirn. The other end of the bobbin or the pirn has a spring-loaded holder/clamp to prevent it from dislodging during winding. The spindle is driven by the intermediate shaft through the friction pulleys. The intermediate shaft is driven by the main shaft which in turn gets the drive from the motor shaft. The thread guide derives motion from the cam through the connecting rods, double rocker arm and the follower. The profile of the cam defines the follower's displacement path which in this case is to and fro. This to and fro motion of the follower is transmitted to the double rocker arm. The input rocker arm derives motion from the linking rod of the follower. The linear motion of the follower is converted to the oscillatory motion of the input rocker arm by means of a pivot attached to the frame of the machine. The oscillatory motion of the input rocker arm is transmitted to the output rocker arm. The cam derives motion from the driven cone drum through the worm, worm wheel and bevel gears. The driven cone drum derives motion from the main shaft through the driver cone drum. The novelty of the machine lies in obtaining traverse speed and traverse length variation as per the requirement of the package. This is achieved as detailed below.

For reducing traverse speed, the belt connecting the cone drums is shifted by rotating the handle from the position of the larger diameter of the driver cone towards the smaller diameter. The mechanical linkages described above bring about the reduction in the traverse speed of the thread guide. On similar lines, for speed enhancement, the belt connecting the cone drums is shifted by rotating the handle from the position of the smaller diameter of the driver cone towards the larger diameter.

Reduction in traverse length can be achieved by rotating the handle provided for the said purpose in an anti-clockwise direction which results in a change of the position of the pin in the slot of the output rocker arm, thus reducing the oscillating length. An increase in traverse length can be achieved by rotating the handle in the clockwise direction.

The following examples are given by way of illustration of the working of the machine.

Example 1: Winding of carbon tows onto the bobbin.

Bobbin dimensions.

Diameter: 60 mm.

Length: 80 mm (Between flanges).

Flange diameter: 110 mm

Spindle speed: 120 rpm.

Belt position on cone drum—distance from cone drum pulley (12- refer Fig. 46) to belt: 65 mm.

Traverse Speed: 6 strokes/min.

Double rocker arm position—distance between bottom end cap (48- refer Fig. 47) to square flange unit (53- refer Fig. 47): 182 mm.

Traverse Length: 80 mm.

Fixture: Bobbin Holder (79- refer Fig. 50).

Example 2: Winding of carbon tows onto pirn.

Pirn dimensions.

Diameter: 9 mm.

Length: 150 mm (Between flanges).

Flange diameter: 15 mm.

Spindle speed: 120 rpm.

Belt position on cone drum—distance from cone drum pulley (12- refer Fig. 46) to belt: 205 mm.

Traverse speed: 3.2 strokes/min.

Double rocker arm position—distance between bottom end cap (48- refer Fig. 47) to square flange unit (53- refer Fig. 47): 227 mm.

Traverse length: 150 mm.

Fixture:Pirn holder (78- refer Fig. 45).

Acknowledgements The following students and their institutes are acknowledged for their contributions vide their project works. 1. Nithya Aparna B, ME, Design and development of multiple head carbon roving winding machine. UVCE, Bangalore University, 2012. 2. Revanth Y, Ruthresh Kumar R, Shashikanth Mohan and Shivaprasad B V. BE. Mech Engg. The beat-up mechanism for a 3D weaving Machine. MSRIT, VTU 2015-2016. 3. Giridhar Kalkeri, M E in Machine Design. Design of 3D weaving machine. Bangalore University.2014-2015.

Chapter 6
Epilogue

B. S. Sugun

The book has basically attempted to bridge the gap between the laboratory research work and the application requirements of the industry, wherein, the industry needs a sort of ready reckoner to implement the technology. This has been attempted in this book, by discussing the modalities of technology development from the basic conceptualization to its realization. The discussion levels are different for each arterial technology. For example, substantial progress has been made in orthogonal weaving globally, whilst too little developments can be seen in the dual plane shedding domain. Therefore, as is discussed in this book, complex geometrical shapes are being tried out using orthogonal weaving technology, whereas, dual-plane shedding-based 3D weaving is yet to make its mark in the industry, and therefore discusses the machine development basics and the actual building of a prototype encompassing textile loom concepts, machine design concepts and innovative approaches in designing the sub-elements. The angle interlock weaving is at an intermediary level with few applications being explored. Nevertheless, all three technologies are a treatise for the textile technologist who is keen to understand the technical intricacies involved in their practical development. One of the chapters also discusses in-depth the machine design aspects, which may, at the outset, seem to be out of context, but, nevertheless found required to completely describe the efforts that goes into building a textile weaving machine and its sub-elements in a constrained space. It is to be understood that this constraint is not an imposed one, but very much inevitable due to the inherent nature of the textile weaving loom and is therefore a real challenge to the mechanical engineer.

At this juncture, it may not be appropriate to make conclusive statements for these technologies, for the sheer reason that, this is just evolving excitedly around the world both on the laboratory research platform and in the textile industries. Few applications

B. S. Sugun (✉)
Centre for Societal Missions and Special Technologies (CSMST), CSIR- National Aerospace Laboratory(NAL), Bangalore, India
e-mail: sugun@nal.res.in

B. S. Sugun (ed.), *Practical Approach to 3D Weaving*, Materials Horizons: From Nature to Nanomaterials, https://doi.org/10.1007/978-3-030-80424-4_6

have trickled in both the structural and thermal sectors of the aircraft industry, which can be sourced from literature. Apart from metals, polymer matrix composites (PMC) are being used in structural members of Aircraft. For 3D preforms to be used in PMC's, there exists the formidable challenge of converting them into composites, especially if the part needs high fibre content. Here, the viscous resin system will have to flow through the dense and complicated weave architecture in a defined time (termed pot life of the resin system). The resin system also has to thouroughly wet the 3D preform without any voids/defects.

On the other hand, the Gas turbine components are focussed towards thermo-structural applications. These composites, termed Ceramic matrix composites, are manufactured by chemical vapour infiltration, which would be advantageous for 3D preforms. Most importantly, the high degree of thermal stresses in the through-thickness direction can be countered by the 3D weave architecture. These composites, do not need high fibre content and hence better prospects for 3D woven preforms exists here. To start with, for 3D preforms and their composites, the focus would be in the strategic sector. As is the trend, once proven in the strategic sector, such technologies percolate downstream to the automobile and construction industry. Also, the medical sector will not be far off and a different bundle of applications can be expected. For the creative mind, scope of application exists even in the fashion industry. Specific applications have not been discussed here, as the reader can get this information from the literature.